Experimental Economics

Experimental Economics
Volume II: Economic Applications

Edited by

Pablo Branas-Garza
Middlesex University London, UK

and

Antonio Cabrales
University College London, UK

Translated by Manuel Muñoz-Herrera
University of Groningen, The Netherlands

Technical review by David Kernohan
Middlesex University London, UK

To María

To Isabel

Contents

List of Figures

List of Tables

Acknowledgments

This book starts with the usual thanks to our collaborators, which sounds commonplace, but in this case is heartfelt. This book is the fruit of a collective effort, gathering the work of researchers from Barcelona, Berlin, Bilbao, Bogotá, Calgary, Castellón, Düsseldorf, Granada, Guatemala, Madrid, London, Los Angeles, Malaga, Milan, Norwich, Reading, Rome, San José, Valencia and Virginia.

We can also say that in this book "you can find (almost) all of those who should be found" – very few are missing! – and, without a doubt, "all those who are found should be found." As our dear friend Nikos would say, in this book there is a lot of *American Economic Review, Econometrica, Games & Economic Behavior, Journal of Political Economy, Quarterly Journal of Economics,* etc…, there is even *Science, Proceedings B* and *PNAS*.

We thank all those who contributed – we have a spectacular book! We also thank Antoni Bosch (Universitat Pompeu Fabra), because he was the seed from which the idea for this book germinated. Manuel Muñoz-Herrera and Sharlane Scheepers translated the entire book from the original Spanish. Akash Sharma dealt with the tables, figures and more besides; David Kernohan revised the English version; Balint Lenkei helped a lot; Middlesex University London and University College London provided the means, both material and human, to make the project possible.

We would also like to thank all of our doctoral students. While we will not say so to our deans, they are much more of a benefit than a cost and they have taught us many things that have helped us grow. Pablo thanks Ramón, Natalia, Luismi, Patricia, Máximo, Segun, Antonis, Filippos, Espín, Marisa and Levent. Antonio thanks Nicola, Matthias, Domenico, Neus, Alejandro, Óscar, Paolo, Toni, Margherita, Ana Paula, Flavia, Sandro, Javier, Sevinc, Jesper, Daniel, María and Nikolas.

We dedicate this book to those who put up with us at home.

Pablo: I thank Maria for her immense patience and the respect she has for my work. Over the past few years she has endured and has been even more patient with my absence, schedules, and much more.

On top of all of this, there was us moving to London.... We have been in Finchley for three years! This book has also taken me away from my three daughters, Paula, Marta and Ana. I dedicate this book to them.

Antonio: I dedicate this book to my wife, Isabel, who includes among her many virtues that of being amused by a husband who is at sea in strategic situations, while trying to research and teach them. Even when I change my mind about what I want to be (or where I want to live) when I grow up, she continues to be amused. And to my children, Ana and Pablo, who have shown that teenagers can be nice, smart, hardworking, charming and understanding of their parents.

Notes on Contributors

Francisco Alpízar is the Director and PI of the program EfD Initiative at the Centro Agronómico Tropical de Investigación y Enseñanza (CATIE), Costa Rica, and Deputy Director of the Latin American and Caribbean Environmental Economics Program. He is also an associate faculty member of Goteborg School of Economics, Sweden. He completed his degree in Economics at the University of Costa Rica (1996) and holds a PhD in Economics from Goteborg University (2002). His articles have appeared in leading journals, such as the *Journal of Public Economics, Environment and Development Economics, World Development, The Scandinavian Journal of Economics, Economics Letters* and *Ecological Economics*. He is an associate editor of *World Development*.

Giuseppe Attanasi is tenured Assistant Professor of Economics and Director of the Laboratory of Experimental Economics at University of Strasbourg, France. Previously he was Junior Chair in Mathematical Economics at Toulouse School of Economics. He completed his PhD in Economics in 2006 at Bocconi University of Milan. His research focuses on behavioral decision-making, game theory and experimental economics. His work has appeared in *International Game Theory Review, Journal of Behavioral and Experimental Economics, Journal of Economics Psychology, Management Science, Marketing Letters, Organizational Behavior and Human Decision Processes, Pacific Economic Review,* and *Theory and Decision*.

Diego Aycinena is Catedrático at the Facultad de Ciencias Económicas and Director of the Centro Vernon Smith de Economía Experimental at Universidad Francisco Marroquín, Guatemala and a research affiliate at the Economic Science Institute at Chapman University, California. He holds a PhD from George Mason University. His research interests include experimental economics, economic systems design, behavioral economics and field experiments. His articles have appeared in *Experimental Economics, The Review of Economics and Statistics, American Economic Journal: Applied Economics,* and *PLOS ONE*.

Antoni Bosch-Domènech is Professor of Economics, Director of LeeX (experimental economics lab) and, is a vice-rector for finance and strategic projects at Universitat Pompeu Fabra in Barcelona, where he has also been Chair of the Department of Economics and Business. His articles have appeared in *The American Economic Review, Games and Economic Behavior, Economic Journal, Experimental Economics*, and *Journal of Risk and Uncertainty*.

Jordi Brandts holds a PhD in Economics from the University of Pennsylvania. He is a research professor at the Institute for Economic Analysis (CSIC) in Barcelona. He is also affiliated with the Barcelona Graduate School of Economics and a research fellow of CESifo. His research is experimental in areas such as the study of cooperation, organizational economics, industrial, organizational and market analysis, conflict and the effects of communication on strategic inter-action. From 2008 to 2013 he held the Serra-Ramoneda/Catalunya Caixa Chair in the Department of Business of the Universitat Autonoma de Barcelona. His articles have appeared in *The American Economic Review, The Economic Journal, Management Science* and *The Journal of the European Economic Association*. From 2007 to 2011 he was editor-in-chief of the journal *Experimental Economics*. Since 2013 he has served as advisory editor for *Games and Economic Behavior*.

Juan Camilo Cárdenas is Professor of Economics at the Universidad de los Andes in Bogotá, Colombia since 2004. He completed his PhD in Resource Economics at the University of Massachusetts Amherst in 1999 and conducted his postdoctoral work at the Elinor and Vincent Ostrom Workshop in Political Theory and Policy Analysis in 2000. His research focuses on the study and design of institutions that promote cooperation among individuals in the most efficient, fair, democratic and sustainable possible way. His research has appeared in journals such as *Science, Journal of Development Economics, World Development, Ecological Economics, Land Economics, Agricultural Systems, American Economic Journal: Applied Economics, Journal of Economic Behavior & Organization* and *Environmental and Resource Economics*.

Enrique Fatas is Professor of Economics at the University of East Anglia and a co-investigator at the ESRC Network for Integrated Behavioural Sciences (NIBS). His interdisciplinary research deals with human behavior. His work has been published in several journals in Economics and other disciplines (including the *Proceedings of the*

National Academy of Sciences, Management Science, and *Psychological Science*). His research areas are behavioral economics, public economics, organizational behavior, industrial organization, and the economics of conflict.

Nikolaos Georgantzís is Professor of Behavioural Economics in the School of Agriculture Policy and Development at the University of Reading, UK. Previously, he was a professor at the Universitat Jaume I and the University of Granada. He studied Economics at the University of Piraeus and has an MPhil from UC Swansea and an MA and PhD from the European University Institute, Florence, Italy. His research interests range over a broad spectrum of topics, including economic psychology, industrial organization, behavioural economics, labour economics, environmental economics, and international economics. His articles have appeared in journals such as *Frontiers in Behavioral Neuroscience*, and *Organizational Behavior and Human Decision Processes*. He is an associate editor of the *Journal of Behavioral and Experimental Economics* and *PLOS ONE*.

Hubert János Kiss is an assistant professor at Eötvös Loránd University, Budapest, Hungary since September 2011. He is also a research fellow in the Momentum (LD-004/2010) Game Theory Research Group at the MTA KRTK, Hungary. Previously he was a visiting professor at the Universidad Autónoma de Madrid. He completed his PhD in Quantitative Economics in 2009 at the University of Alicante. His research focuses on financial stability with special emphasis on bank runs and on behavioral and experimental economics. His articles have appeared in *Journal of Economic Behavior & Organization, Journal of Money, Credit and Banking, Journal of Financial Stability, Southern Economic Journal*, and *Journal of Behavioral* and *Experimental Economics*.

Francisco Lagos is Associate Professor of Economics at the University of Granada, Spain since October 1999. Previously he was Assistant Professor of Economic Theory at the University of Alicante. He completed his PhD in Quantitative Economics in 2003 at the University of Alicante. His research focuses on experimental games and economic behavior, with special emphasis on labor markets. His articles have appeared in *The American Economic Review, Games* *and Economic Behavior, Human Brain Mapping, Economic Inquiry, Social Choice and Welfare, Economica* and the *Journal of Population Economics*.

Humberto Llavador is Associate Professor of Economics at the Universitat Pompeu Fabra, Spain and an affiliate professor of the Barcelona Graduate School of Economics. He was a member of the school of Social Sciences at the Institute for Advanced Study in Princeton in 2009, and has held research and teaching appointments at Yale and the London School of Economics. His research focuses on political economy, climate change and welfare economics, and his studies have appeared in *The Quarterly Journal of Economics*, the *Journal of Public Economics*, the *Journal of Development Economics and Climatic Change*, among others. He is the author of the book *Sustainability for a Warming Planet* (2015) on the economics of climate change. In 2012 he received the Recognition Jaume Vicens-Vives for teaching quality and innovation.

Debrah Meloso is Associate Professor of Finance at the ESC-Rennes School of Business, France since August 2014. Previously she was a professor in the Department of Decision Sciences at Bocconi University. She completed her PhD in Social Sciences, specializing in Finance at Caltech in 2007. Her research focuses on the validity of various notions of equilibrium in laboratory settings, mainly notions related to General Equilibrium models. Her articles have appeared in journals such as *Science* and *Management Science*.

Antonio J. Morales is Associate Professor of Economics at Universidad de Málaga, Spain since 2003. He holds a PhD in Economics from the University College London. His research focuses on behavioral and experimental game theory, with special emphasis on conflict behavior, public goods and models of learning and bounded rationality. His articles have appeared in *Econometrica, Journal of Economic Behaviour & Organization, Journal of Risk and Uncertainty, Applied Economic Perspectives and Policy, The B.E. Journal of Theoretical Economics, Southern Economic Journal*, and *Journal of Economic Psychology*.

Robert Oxoby is Professor of Economics at the University of Calgary, Canada. He holds a PhD from the University of California, Davis. He is a senior fellow of the Canadian Institute for Advanced Research (Social Interactions, Identity, and Well-Being working group) and a research fellow with the IZA Institute for Labor Research. His research focuses on topics in behavioral and experimental economics, with special interests on topics of regarding how social and personal identity feedback on market institutions. His research has appeared in *The Economic Journal, The Quarterly Journal of Economics, Personality and Individual Differences*, and the *Journal of Applied Social Psychology*.

José Penalva is an associate professor at the Universidad Carlos III in Madrid where he teaches the PhD and the Master in Finance programmes, as well as at the undergraduate level. He holds a PhD in Economics from UCLA. He is currently working on information models and finance market microstructure. His research has appeared in *Econometrica, The B.E. Journal of Theoretical Economics*, and the *Review of Economic Dynamics*.

David Porter is Professor of Economics and Mathematics, and the Donna and David Janes Endowed Chair in Experimental Economics at Chapman University. He obtained his PhD in 1987 from the University of Arizona. His research interests include economic systems design, financial economics and experimental methods. His work has appeared in the *Proceedings of the National Academy of Sciences, The American Economic Review, The Economic Journal, Journal of Economic Behavior & Organization, Review of Finance, Experimental Economics, Journal of Behavioral Finance*, and *the American Economic Journal: Microeconomics*.

Stephen J. Rassenti is Professor of Economics and Mathematics, and the Director of the Economic Science Institute at Chapman University, California. He holds a PhD in Systems Engineering from the University of Arizona in 1981. His research interests include economic systems design, experimental economics and organizational design. His work has appeared in the *Proceedings of the National Academy of Science, The American Economic Review, Science, Annals of Operations Research, RAND Journal of Economics, Bell Journal of Economics, Games and Economic Behavior, Econometrica, Experimental Economics, The Economic Journal*, and *Journal of Economic Behavior & Organization*.

Ernesto Reuben is Assistant Professor of Strategy at the Columbia Business School, New York. His main research interests lie within behavioral economics. In particular, on the determinants of prosocial and antisocial behavior, the emergence and enforcement of social norms, the influence of interest politics, and the role of behavioral biases on labor market discrimination. His work has appeared in outlets such as the *Proceedings of the National Academy of Sciences, American Journal of Political Science, The Economic Journal, Journal of Public Economics,* and *Games and Economic Behavior.* He is currently an associate editor of the *Journal of Economic Behavior & Organization.*

Ismael Rodriguez-Lara earned a PhD in Quantitative Economics from the University of Alicante in 2010. Previously he was Lecturer in Economics at Universidad de Valencia and a research fellow at LUISS Guido Carli University, Rome. Since September 2013, Ismael works as Senior Lecturer in Economics at Middlesex University London. His research focuses on game theory, behavioral finance and experimental economics. His main areas of expertise are fairness ideals, principal–agent relationships, and coordination problems. His articles have appeared in *Journal of Money Credit and Banking, Journal of Economic Behavior & Organization, Experimental Economics,* and *Journal of Behavioral and Experimental Economics.*

Alfonso Rosa-Garcia is an assistant professor at the Universidad Catolica San Antonio de Murcia, Spain since September 2012. He completed his PhD in Quantitative Economics in 2012 at the University of Alicante. Previously he was Lecturer in Economics at the Universidad de Murcia. His research focuses on network economics, behavioral finance, experimental economics and teaching of economics. His articles have appeared in journals such as *Journal of Money Credit and Banking, Journal of Economic Behavior & Organization, Journal of Behavioral and Experimental Economics and International Journal of the Commons.*

Joaquim Silvestre holds a PhD in Economics from the University of Minnesota in 1973. He is Professor of Economics at the University of California, Davis. Previously he taught at the Universitat Autònoma de Barcelona from 1973 to 1983 and visited UC Berkeley, UC San Diego, the Institut d'Anàlisi Econòmica and the Universitat Pompeu Fabra. His research articles on economic theory, public economics and experimental economics have appeared in journals such as *Econometrica, Journal of Economic Literature, Journal of Economic Theory, European Economic Review, The Review of Economic Studies* and *The Economic Journal.* His recent books include *Public Microeconomics* and *Sustainability for a Warming Planet* (with H. Llavador and J. E. Roemer).

Carles Solà Belda is an associate professor in Departament d'Empresa, Universitat Autònoma de Barcelona since 2008. His research focuses on experimental methods with applications to organizational behavior and personnel economics. His main interest lies in the interaction between individual motivations and organizational objectives and outcomes in recruiting, promotions, leadership behavior, and compensation systems, among others. His articles have appeared in *Games and Economic Behavior* and *Journal of Economic Behavior & Organization*.

1
Market Organization and Competitive Equilibrium

Antoni Bosch-Domènech and Joaquim Silvestre

Introduction

The market (or better, markets) is any system that facilitates exchange. It is therefore a necessary condition for economic activity, as well as the subject matter of economic science. The theoretical analysis of markets, pioneered by classical economists (Walras, 1874, Edgeworth, 1881) and continued by Marshall (1890), culminates in the precise model of perfect competition (Arrow and Debreu, 1954). The competitive model postulates that participants in the market decide the quantities that they wish to buy or sell according to the market price, which each participant takes as given. A price is an *equilibrium price* if the buying and selling plans of the various participants are compatible.

The competitive model is static: it determines the equilibrium price and quantity, but it does not specify how to reach them. In a way, it disregards (ephemeral) disequilibrium situations. When forced to provide some justification for the dynamic path to equilibrium, Walras appeals to the metaphor (understood as such by both Walras and his successors) of a virtual auctioneer, who adjusts prices according to the difference between supply and demand, and does not allow trade to occur until the equilibrium between supply and demand is attained. But in fact the competitive equilibrium model is silent on the manner in which transactions are performed and how equilibrium is arrived at.

Wanna trade?

The experimental study of competitive markets begins with Edward Chamberlin (1948) and matures with the experiments of Vernon Smith (1962, 1964). Both attempt to test the competitive model in an isolated market (i.e., to test whether the theoretical values of the competitive equilibrium model are good predictors of the magnitudes reached by prices and quantities in the experiment) as well as to check whether, as implied by the competitive model, the experimental market outcome maximizes the sum of the profits for buyers and sellers, in which case we say that we have reached 100% efficiency. Yet market experiments allow us to observe the buying and selling decisions made by the experimental participants (or "subjects") *in real time*: as a result, we are induced to focus on the dynamics by which individual decisions drive economic variables, sometimes towards equilibrium and at other times away from it.

In order to construct an environment, no matter how simple, in which to explore the behavior of a market, one must establish the rules by which the market should operate (i.e., the norms that regulate how participants can bargain, how agreement is reached, or how to ensure that agreements are implemented). No market can exist without operating rules, be they simple or complex, explicit or implicit, because a market is essentially the set of its operating rules. For this reason, when an economist refers to a market she means the rules and institutions that define it, rather than the physical location where trade occurs (say, the village market place, or the central market in Southampton).

Two important features differentiate the experimental market rules of Chamberlin from those of Vernon Smith. First, in Chamberlin's experiment participants move about a room and try to reach a buy–sell agreement with another participant. As in more primitive markets, sellers and buyers are scattered in space and time, and this entails a transaction cost because they have to find each other and negotiate one on one. Vernon Smith (1962) replaces the rule of one-on-one bargaining with a more modern institution: he creates an environment where buyers and sellers meet by rules that publicly disclose their offers to buy or sell in real time, as well as the prices of the trades already contracted. More specifically, it is a form of auction, called a *double auction*, where buyers increase their buying (*bid*) prices and sellers decrease their selling (*ask*) prices until some bid price coincides with an ask price and a deal is closed. Hence, the main difference between Chamberlin's and Vernon Smith's markets is that participants in the former have to search for trading partners and bilaterally negotiate, whereas in the latter negotiation is multilateral: everybody is aware of what everybody else is doing. The *information* on the various bids and asks, as well as on the prices of the transactions reached, becomes a public good: both buyers and sellers may use the information at zero cost.

The second difference is that, whereas Chamberlin ends the experiment once all possible transactions have occurred, Vernon Smith repeats the experiments several times, keeping the same parameters, but without carrying over the outcomes of any one round. Each round is independent of the other rounds: the experiment starts each time at zero, but of course the participants acquire experience that will no doubt influence their decisions in the following rounds. What is the motivation for repeating the experiment?[1] Vernon Smith explains that, if we want to test whether the market converges to the theoretical competitive equilibrium, we must allow participants in the experiment to acquire experience and modify their behavior in light of previous experience, as happens in real-life markets. Rather than trying to confirm that prices coincide with competitive equilibrium prices from the first transaction, the aim is to check whether there is a trend of convergence towards the equilibrium, and, if so, what is the speed of convergence.

But the preferences of the participants are similarly "induced" in either experiment: each buyer and each seller is told her (or his) *reservation price* for buying or selling one indivisible unit of the (fictitious) good traded in the experimental market.[2] When we, the experimenters, fix the reservation price of each participant (i.e., for a buyer, the maximum price at which she will be willing to buy and, in the case of a seller, the minimum price at which she will be willing to sell), we induce in each participant "her" preferences which, obviously, may well vary from person to person.[3,4] To summarize, the reservation price is private information for each participant: only she (and of course, the experimenter) knows it.

Note that, because the experimenter knows all reservation prices (she has provided them), in order to construct the demand curve for this market, all she has to do is to aggregate the buyers' reservation prices. Similarly, by aggregating the sellers' reservation prices she will obtain the supply curve.[5] Once she has determined the supply and demand curves, it is trivial to find the competitive equilibrium price and quantity from the point of intersection of the curves. Next, after knowing the equilibrium values, she can compare the theoretical values with the prices and quantities observed throughout the experiment and, using this comparison, she can ascertain the extent to which the equilibrium of the competitive model provides a good approximation to the experimental results observed. This last sentence should be emphasized because it encapsulates the capacity of an experiment to act as a method to test, or check, a theory: the idea is to create an environment in which the predictions of the theoretical model are precise, so that one can experimentally verify the extent to which they are fulfilled.

Finally, we should note that the incentives of the participants in either experiment were similar. In each period, sellers could sell a maximum amount of the "good," and buyers could only buy a given quantity of the "good."[6] A buyer's earnings were higher the greater the difference between her reservation price and the price paid for the purchase, and a seller's earnings were higher the greater the difference between the price received in the sale and her reservation price.

Why, at the risk of being tedious, do we go to such lengths to explain the details of the rules imposed on their markets by Chamberlin and Vernon Smith? First, because their designs have been copied by many subsequent experiments.[7] Second, because, as we discuss below, we face a crucial issue in experimental methodology. Chamberlin observed that the predictions of the model were not realized, and interpreted his results as supporting the hypothesis that what happens in real markets differs from the predictions of the theoretical competitive model. Vernon Smith, on the contrary, observed a very clear tendency to converge towards the predictions of the competitive model, and, moreover, that the convergence was fast. In addition, the experimental markets approached the 100% efficiency predicted by the competitive model. Leaving aside for the moment a detailed analysis of these results, we must make a key methodological remark. We have seen how a change in the design of the experiment, which at first blush may not seem essential, may drastically alter the outcome of an experiment. Moving from individual bargaining to a market organized by a double auction allows us to validate a hypothesis that had been disproved with the first design. The reader should draw an important lesson from this: details are very important in the design of an experiment. *Sometimes a small alteration can substantially modify the results of an experiment. Thus, one may ask, how can you trust an experimental result? The answer is obvious but important.* One must *repeat and replicate* the experiment *in order to test its robustness to variations.*

Makin' magic

The miracle

Let us now pay attention to the results of the experiments described. Few people are surprised when the predictions of a theoretical model are not fulfilled in reality; perhaps because of this, Chamberlin's experiment went relatively unnoticed.[8] Yet it is nothing short of extraordinary that the competitive model, which hypothesizes a price known to all market participants but not manipulated by any of them, can be an excellent predictor of what happens in an experimental market, where these and other assumptions (presented in detail below) are not met.

But is the competitive model such a good predictor of Vernon Smith's experiments? Consider Figure 1.1, which is representative of thousands of similar experiments conducted over the years. On the

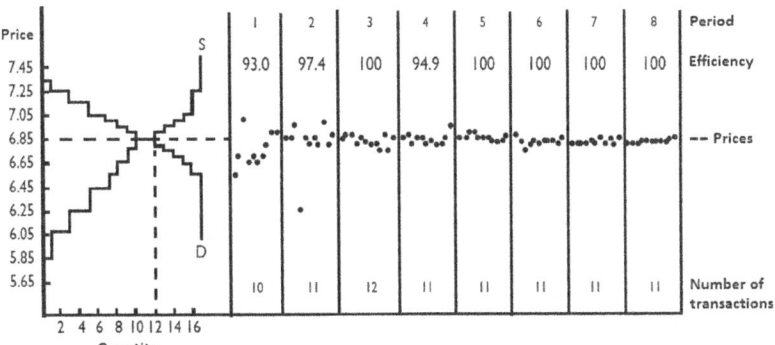

Figure 1.1 Vernon L. Smith (1962)

Note: Equilibrium price and quantity. Experimental prices and quantities across the eight periods of the experiment.

left are the curves of supply and demand that the experimenter knows (but as will be recalled, are not known by the participants in the experiment) and allow the experimenter to compute the price and the equilibrium quantity. On the right, represented by dots, we have the prices at which each transaction has been carried out in each of the periods of the experiment: eight in this case, numbered from one to eight; the vertical lines are used to separate periods. For each period, under the period number, we read the percentage of market efficiency in the period (93.0, 97.4 . . .) and, at the bottom, another number indicates the number of transactions that have occurred in the period (10, 11, 12, 11, . . .). What do we see in the graph? At least three things: a rapid convergence of prices towards the equilibrium price (6.85) and of efficiency towards 100%, as well as transaction volumes that stabilize in the middle of the three equilibrium quantities – 10, 11 and 12.

We do not know what your reaction was when you first encountered this result. As noted, the competitive model is based on extreme assumptions. In detail:

(a) Buyers and sellers take the price as given.
(b) The above assumption is justified by the idea that buyers and sellers are so numerous, and each of them is so unimportant that none of them in isolation can have an effect on prices: for example, both Cournot (1838) and Edgeworth (1881) justify the above assumption as a limit when the number of participants tends to infinity and the weight of each of them in the economy becomes negligible.[9]
(c) Participants know the market price.
(d) Knowing the price, agents make decisions on how much they want to buy or sell in order to maximize their own profit or utility.

Well, perhaps in the experiment participants did make maximizing decisions (although we cannot be sure of that). Yet we certainly can be sure that buyers and sellers in the experiment in no way felt that the price was fixed, and not only did their best to manipulate the buying and selling prices, but actually managed to do this. Nevertheless, the predictions of the model are fulfilled. Furthermore, the experimental results were obtained without the participants needing to know the "market price:" they just needed their private information and the *information* acquired by observing the various bids and offers, as well as the actual prices of a few transactions in a few repetitions. (This is particularly the case in Figure 1.1's experiment, where 100% efficiency is reached and maintained after five repetitions. After that, prices just fluctuate around the equilibrium value and the quantities exchanged are the equilibrium ones from the beginning.) Finally, the convergence of the prices and quantities to the competitive equilibrium is achieved with only four or five buyers and four or five sellers: far from the infinite numbers proposed by Cournot and Edgeworth.

Although we do not know what your reaction was, we do know Vernon Smith's reaction because he put it in writing: *"I am still recovering from the shock of the experimental results. The outcome was unbelievably consistent with competitive price theory. . . . But the result can't be believed, I thought. It must be an accident, so I will take another class and do a new experiment with different supply and demand schedules."* (Vernon Smith, 1991, p. 156.)

We can say today that the miracle and its mystery survive after all these years. What is going on during the process of proposing to purchase and sale prices that leads the double auction market to converge, efficiently, to the equilibrium price and quantity? Buyers and sellers know nothing about the equilibrium price: they do not even know whether their purchase or sales proposals can be met. Yet the process converges rapidly towards equilibrium. This discovery has without doubt changed the way experimentalists understand the economy and markets.

The load test

As Vernon Smith did when he encountered his result, we experimentalists must keep a certain degree of skepticism about any experimental result, whether surprising or not (particularly if it is not surprising).

We must always ask ourselves whether the result depends on the specific features of the experiment, in which case it would be unwise to extrapolate its results. And in order to address this question we must replicate the experiment until we find the limits beyond which the results initially observed no longer obtain. "Let us change the supply and demand," proposes Vernon Smith, and see if anything happens. Well, decades later we know what has happened when the supply and demand have been changed thousands of times over the years: the results have held. The miracle is true! The predictions of the competitive model are correct when the institution of exchange is the double auction, regardless of the supply and demand curves.[10]

But these results depend on factors other than the specific parameters of the supply and demand curves. To begin with, participants in these experiments are usually students; and in many cases, students of economics. Are we sure that the experimental results would hold with other types of participants? No, we are not sure. Questions of this kind can only be answered empirically: we do not know until we test them. But again, after decades of trial and error with students of all levels, with businessmen, with ordinary people in countries as diverse as China, Saudi Arabia and Paraguay, with teachers, with civil servants, . . . the result holds: it is robust.

The euphoria seems justified, but we should be able to contain it. As noted above, an important lesson from these experiments is that sometimes small changes in the rules of the game affect the results. We should therefore ask what changes in the rules or institutions may affect the results, and to what extent. What if the sale and purchase agreements, rather than being implemented by the method of double auction, are executed according to the procedure, usual in many of the markets of our countries, where each seller *sets* the price at which she wants to sell her commodity, and the buyer decides whether to buy at any of the prices set by different sellers? Well, in the case of experiments in which each seller announces her price, which is non-negotiable (*posted price*), some convergence towards equilibrium prices and quantities and to the maximum efficiency is usually observed, but the convergence is *slower* and *less precise* than in the case of the double auction. See, for some early experiments with fixed prices, Plott and Smith (1978), as well as Hong and Plott (1982), who highlight the inefficiencies of design in the river and canal transportation markets in the United States. Similar conclusions are reached when using other types of rules, such as *sealed auctions*, in which buyers and sellers send their purchase prices and sales to an "auctioneer" who implements the transactions that are possible at the prices received (see, for example, Kagel and Levin, 1985).

In conclusion, we observe that experimental markets work better (double auction *for example*) or worse (posted prices *for example*) depending on the rules by which they operate. Some operating rules result in very efficient markets, while other rules lead to less efficient markets.

The scare

After observing and confirming the miracle, a natural question is: "Did the participants in these experiments do something special to get the market to converge quickly to the prediction of the competitive model?" It seems clear that some element of rationality by subjects is indispensable for obtaining these results. Right?

Wrong! Gode and Sunder (1993a, b) showed that a good institution (a good rule) can compensate for the presence of foolish participants, even when all participants are foolish. In order to test the hypothesis, they created computerized robots with zero intelligence, separated into buyer and seller robots. A seller robot was endowed with one unit to be sold, which unit had a selling reservation price, and all the robot could do was to *randomly* propose a selling price not lower than its reservation price. In a parallel manner, a buyer robot had a buying reservation price, and all the robot could do was to *randomly* propose a buying price not higher than its reservation price.[11] Therefore, the robots were neither maximizing profits, nor could they learn from what was happening in the market. A robot just chose its random price when its turn arrived. If the purchase price (chosen at random by a buyer robot) was higher than the selling price (chosen at random by a seller robot), then a transaction would take place between the two robots at an intermediate price.

What do we observe in this market populated with foolish robots? It turns out that prices and quantities converge to the competitive equilibrium, and market efficiency is close to maximum efficiency. Therefore, it suffices that no robot is allowed to suffer losses to fulfill the predictions of the model. Put another way, if the irrationality of losing money is not allowed, the double auction market is able to correct the irrationality of some participants who make decisions at random, therefore keeping the trading process near equilibrium (see Figure 1.2).[12]

If a market populated by robots that make decisions at random is efficient, are we experimental economists the mediums who, by our manipulations, make visible the ectoplasm of the invisible hand that guides the market toward its culmination?[13] Furthermore, even though the result of Gode and Sunder corresponds to a simple market with a single good, it appears that it can be generalized to more complex markets.[14] Some institutions can substitute some of

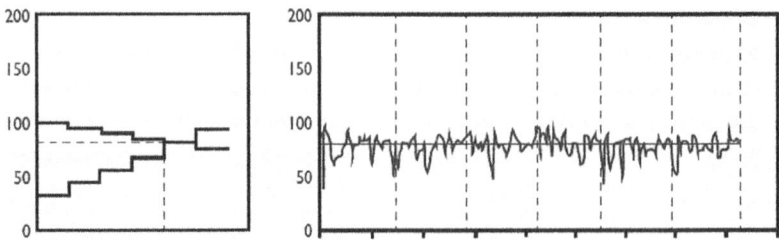

Figure 1.2   Gode and Sunder (1993a, b)

Note: Note that prices fluctuate around their equilibrium value, but they converge to their equilibrium value at the end of each period (marked by the vertical lines). What we do not see in this market, as we shouldn't, is robots learning from one period to another.

the intelligence that helps buyers and sellers find the equilibrium in a market.

The beyond

In order to understand the "beyond" we must be sure we understand the "within." What cases have we seen so far? We have considered the case of a *single* market (because only one good is exchanged) involving a *small number* of experimental participants, with private valuations (maximum buying price, or reservation price, for buyers, and minimum selling price, or reservation price, for sellers), where sellers obtain profits from their decisions (the difference between the reservation price and the transaction price) with the market governed by pre-established rules. Of all the elements that define the experiments described, we discussed the result of changing the rules of the game (double auction, one-on-one bargaining, posted price, . . .), of changing the parameters of private valuations (which implies changes in the supply and demand curves), and even of replacing human subjects with foolish robots that randomly choose their buying or selling prices.

If this was the "within," the "beyond" should be to vary the elements that have remained constant throughout these experiments. Consider, more precisely, a market where it is not the case that each participant has her own private valuation of the good to be traded, where an asset is traded, the value of which is in principle the same for everyone but is not known with precision by any participant. The subjective evaluation of the asset by a participant may then rely on inaccurate information signals or even on the subjective assessment of other participants. This may occur in actual asset markets; for instance, in the stock exchange. If the good is an asset that, besides having a market value, yields its owner a return – for instance a corporation share that yields dividend income – this market is easy prey for speculation, irrational exuberance and mass psychology (discussed in Chapter 4 below).[15] Another example is provided by a mining concession allocated in a sealed-bid auction, where the "winner's curse" is a frequent occurrence, as explained in Chapter 3 below.

By the same token, we may consider markets where a participant has *market power*, in the sense of a greater ability to influence market outcomes than other participants, in which case we can no longer speak of a competitive market. This would occur, in particular, if there were only one seller facing many buyers (in which case we could refer to a *monopoly*), or if there were only two sellers (corresponding to a *duopoly*). We know from economic theory that the predictions of monopolistic and duopolistic models differ from the predictions of the competitive model. Therefore it is natural to ask under what circumstances these different predictions are verified. We do not deal with this issue in the present chapter: the reader is referred to Chapter 2 below.

General equilibrium

So far we have considered single-market experiments. But we know that, in the real world, markets are inter-related and operate simultaneously. We know, for example, that what happens in the labor market affects the goods market: if no one is working, to take an extreme example, and labor is the only source of income (no previously accumulated savings), no one can buy anything in the goods market. In fact, the competitive equilibrium model of Walras, Arrow and Debreu explicitly considers the interaction among a possibly large number of markets. Finding the competitive equilibrium values for all markets, what we call the *general competitive equilibrium*, then requires solving a system of simultaneous equations.

Goodfellow and Plott (1990) were the first to experimentally test a model of general equilibrium. In their experiment, all participants were able to buy and sell in *two* markets that were governed by the rules of the double auction. One could say that, in one market, entrepreneurs were buying labor sold by workers, whereas in the other market entrepreneurs were selling to the workers, now acting as consumers, the goods that they had produced with the labor bought in the first market. As seen in Figure 1.3, the outcomes were consistent with the general competitive equilibrium values.

As we know, the competitive general equilibrium model predicts the full and efficient use of all resources, but in reality market economies experience episodes of low activity. One possible explanation for these phenomena, which may be relevant today, is the existence of credit constraints. Bosch-Domènech and Silvestre (1997) experimentally explore the consequences of introducing credit constraints into the design of Goodfellow and Plott. A general equilibrium model modified to capture credit constraints predicts two regimes: a high credit regime where additional credit availability has no effect on the economy; and a low credit regime, where the availability of credit has effects on both economic activity and on prices. These predictions are supported by the experimental results. Figure 1.4 shows the experimentally obtained prices of the inputs, and compares them with their theoretical equilibrium values.

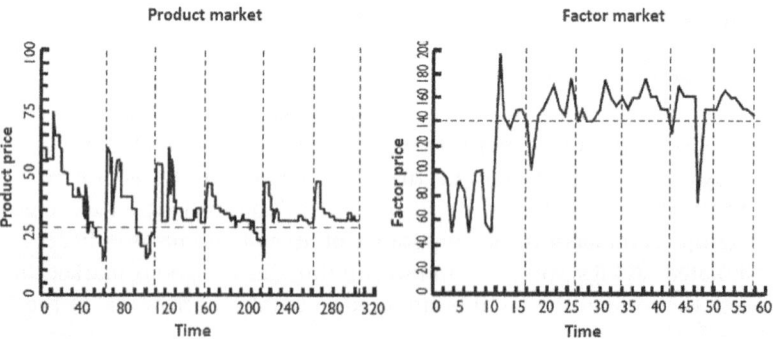

Figure 1.3   Goodfellow and Plott (1990)

Nots: Consider three goods (input, output and numeraire) that are simultaneously exchanged in two markets. The outcomes of a period in an experiment are not carried over to subsequent periods. The vertical dotted lines separate the different periods, while the horizontal dotted lines correspond to the general competitive equilibrium prices. The figure compares the prices obtained experimentally with the theoretical values, and shows a considerable degree of convergence of the former to the latter.

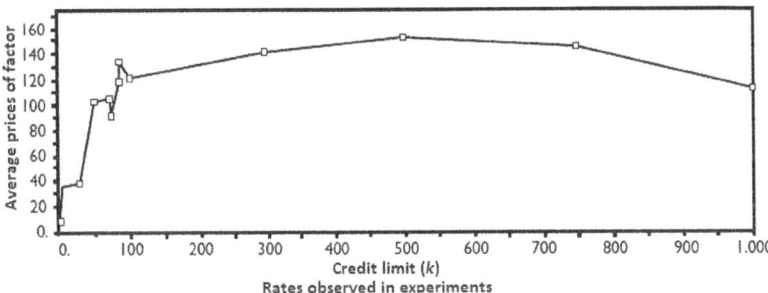

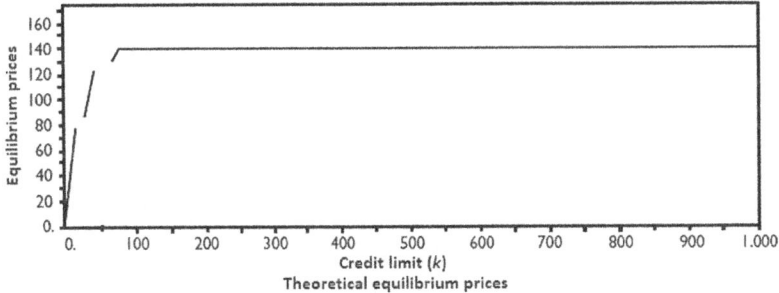

Figure 1.4   Bosch-Domènech and Silvestre (1997)

Notes: This experiment also has three goods (input, output and numeraire) which are simultaneously traded in two markets, but a credit constraint is introduced: the parameter k indicates the degree to which future wealth can be used to fund current consumption. A high value of k reflects perfect creditworthiness, as postulated in the competitive general equilibrium model, while $k = 0$ indicates zero credit. The figure displays the input prices predicted by the theoretical model as well as the prices obtained experimentally as functions of the credit parameter k. The theoretical model predicts (lower panel) that k has no effect on the economy for values above a critical value of k, but it does have effects for k below the critical value: the experimental results (top panel) are consistent with this prediction.

But single-period models are not suitable for studying the role of fiat money. Two papers, namely Lian and Plott (1998) and Hey and di Cagno (1998), consider economies that last for several periods, with two goods in each period and money serving as a link among the periods. These are relatively complex economies, with stocks held over periods, bond markets and the possibility of bankruptcy. The purpose of these experiments is to determine whether the predictions of a competitive general equilibrium model of a single period are fulfilled in experiments that extend over several periods. The Lian and Plott experiment finds that relative prices converge better than quantities.

A second objective of the Lian and Plott experiments is to study the impact of a change in the quantity of money: they observe that an increase in the money supply has a positive effect, more or less proportional, in nominal prices but has no influence on real variables (see Figure 1.5). The complexity of this experiment allows Lian and Plott to test several hypotheses. Specifically, Okun's Law (on the negative relationship between changes in the unemployment rate and real GDP) is experimentally verified, but they find no evidence of a Phillips curve, relating inflation to unemployment.

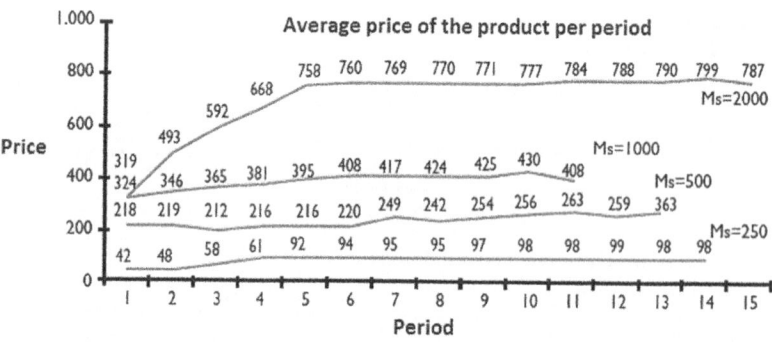

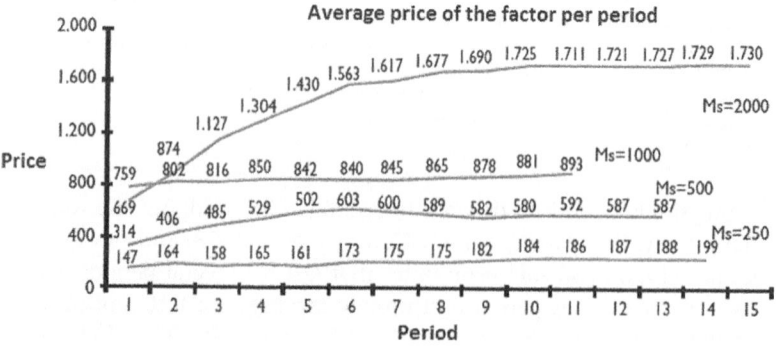

Figure 1.5 Lian and Plott (1998)

Notes: Experimental prices for different levels of the money supply (Ms). After a few periods, prices stabilize at levels roughly proportional to the money supply.

On the other hand, the experiment of Hey and di Cagno aims at verifying the hypothesis that low levels of economic activity are rooted in the sequential nature of markets, by which the workers have to sell their labor before they may express an effective demand for output. Their results fail to confirm that the general competitive equilibrium is reached: the number of transactions is well below the theoretical prediction, efficiency is low, and relative prices are either well above or well below the equilibrium price. In their words "the usual finding that competitive equilibrium is achieved in double auction markets is not replicated in this sequential setting."

The advantage of nations

When trading beyond the borders of a country, domestic conditions in each country tend to influence what goods are imported or exported (i.e., the country's trade specialization pattern). The classical *law of comparative advantage* (Ricardo, 1817) states that a country will tend to specialize in the production and export of those goods for which it has a comparative advantage. This law – in the words of Paul Samuelson, the only economic law that is both real and not trivial – is very difficult to verify with actual data, not only due to the complexity of the factors that influence trade, but also because of the requirement to ascertain that countries export the goods that are relatively cheaper in autarky, and autarky does not exist in the real world.

It is therefore important to put the law to the test in the laboratory. As we know, the law of comparative advantage can be derived from the competitive model. Moreover, from the competitive model one can also obtain another important result in the theory of international trade, called the *factor price equalization theorem*. Noussair, Plott and Riezman (1995) attempted to experimentally verify whether the predictions of the law of comparative advantage and the factor price equalization theorem were fulfilled.[16]

To this end they devised an experiment with two countries, two different kinds of output and a single type of input. The input could not be transferred from one country to another: it was bought by entrepreneurs in the domestic market to be transformed into output, which output could in turn be exported to the other country if there were demand for it. Note that the experimental design is similar to the general equilibrium experiments described above, with the difference that now we have consumers in both countries who compete to purchase the outputs.

The experimental results found by Noussair, Plott and Riezman (1995) clearly confirmed the specialization conjecture posited by the law of comparative advantage. However, the convergence of prices towards competitive equilibrium proved to be slow and erratic. Possibly, when the experimental design is complicated and several markets act sequentially (a need to buy labor first before producing the output), the problems observed by Hey and Di Cagno arise. It may be that employers do not venture to buy labor unless the sale of their output is guaranteed, except when markets are very "deep" in the sense of there being a large potential demand for the product, which may explain why the market moves slowly and erratically.

In Figure 1.6 we can see that prices tend to converge towards the competitive equilibrium, especially in the final periods and, more clearly, at the end of these periods. But the speed of convergence is quite different from that observed in simpler markets. The

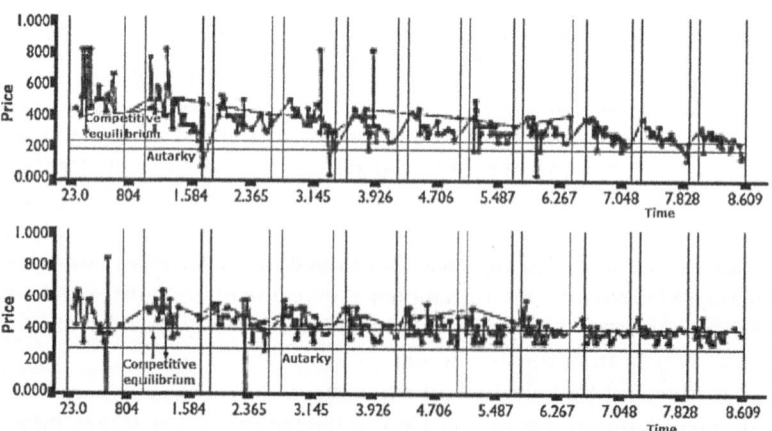

Figure 1.6 Noussair, Plott and Riezman (1995)

Notes: Prices of the two outputs in the two countries. The top panel displays the price of the output of one country, and bottom panel that of the other country.

experiments confirm the pattern of trade specialization by countries, as predicted by the law of comparative advantage.

Last, we refer to a type of large-scale experiment which uses the speed of the internet to allow simultaneous participation of a large number of subjects in different places. Noussair, Plott and Riezman (2007) devised an experiment based on a "globalized" economy in the theoretical model described by about 50 equations. The economy is divided into three countries: each country has its own currency, its own resources and its own technology. There are three final goods that can be produced in any of the three countries. The three countries are endowed with two inputs, both necessary for the production of any of the three final goods. The inputs cannot be sold overseas, but the final goods can. Markets exist for each input, each final good and each currency in each country. Overall, the economy comprises 21 markets which operate simultaneously. The competitive equilibrium of this economy can be found by solving the system of equations, and the solution of these equations determines the set of predictions to be experimentally tested.

This experiment suggests that, despite the complexity of the economy, economic variables move in the direction of competitive equilibrium, and that even though convergence is slower and less complete than in simpler economies, the same principles appear to be operating at both extremes of scale.

Conclusion

In this chapter we have presented probably the most successful topic in the history of experimental economics. No one seems immune to surprise when first seeing the results of the first experiments of Vernon Smith, in which the oral double auction converges quickly to the predictions of the static model of competitive equilibrium, without the need for the theoretical assumptions to be satisfied. This represents a resounding success for the economic theory of perfect competition, which has promoted the study of specific types of auctions that can predict which experimental designs lead to competitive equilibrium outcomes.[17]

Experiments with competitive models have reached, as we have just seen, a significant degree of complexity. Perhaps because of this, core courses in experimental economics rarely refer to experiments with more than one market. Despite constituting a substantial part of experimental research in economics, these lose educational value from their inability to be easily replicated in class. Nevertheless, we have tried here to show the richness of these experiments and the numerous possibilities that they offer for testing a variety of theoretical results. They thus provide a foundation for understanding both micro and macroeconomics, finance and the economics of information, as well as providing a guide to economic policy.[18]

The general competitive equilibrium often spontaneously emerges, "as led by an invisible hand," from the simultaneous operation of a huge number of markets, involving gigantic numbers of transactions among millions of economic agents. Certainly, the design of general equilibrium experiments is not an easy task, but the search for an understanding of the emergence of general competitive equilibrium constitutes one of the most exciting challenges in social science.

Notes

1. The repetition of experiments brings to mind the repetition of days in the movie *Groundhog Day*, with Bill Murray and Andie MacDowell (Columbia Pictures, 1993). It may be recalled that the daily pattern is repeated day after day, and Bill Murray learns from the pattern of repetition to make decisions that better benefit him.
2. The competitive equilibrium model, despite being at the core of economic science since the 19th century, did not seem to lend itself easily to experiments. Researchers could not observe people's preferences or beliefs, or even the production functions of firms. It was the innovation of "inducing" preferences that opened the door to experiments on competitive economies.
3. Hence, the "preferences" of each participant are imposed by the experimenter. Many experiments (mainly in psychology, but also in economics, two disciplines that share a common boundary) do aim at discovering people's preferences (say, towards risk). But in market experiments we impose the participants' preferences in order to specify the theoretical predictions that we wish to test.
4. Even though it falls outside the main topic of the present chapter, it should be recalled that the preferences assigned to each participant are her individual preferences on the outcomes of her actions. If individual and social valuations diverge – as is the case when externalities are present – the valuation of an action by the individual who takes it differs from the social valuation of the action, and the competitive market equilibrium may not be socially efficient. See Theodore Bergstrom and John Miller (1999) for a simple experiment with externalities.

5. See, for instance, Bergstrom and Miller (1999, Chapter 6) for a description of how to obtain supply and demand curves from reservation prices.
6. Each unit of the good could have a different reservation price. Hence, reservation prices could vary not only across people but also by units, in which case the order in which each participant could buy (or sell) each unit was established, starting, for a buyer, with the units with higher reservation prices and, for a seller, with those with lower reservation prices: this reflects the principles of decreasing marginal valuation and increasing marginal cost. The budget constraint simply amounted to the prohibition of buying at a price higher than the reservation price, or selling below it. Unsold units did not affect the profits of sellers.
7. It is instructive to read the first chapter of John Kagel and Alvin Roth (1995) for a description of the impact of the first economic experiments on subsequent ones.
8. This brings to mind the often skeptical use of the expression "in theory."
9. Joseph Ostroy (1980) and Louis Makowski (1980) provide an alternative justification for perfect competition, which in some cases is compatible with low numbers of buyers and sellers.
10. See, for example, the numerous variations in Vernon Smith (1991). The convergence to the competitive equilibrium price is less clear when the demand or supply curves incorporate an element of market power, see Holt, Langan and Villamil (1986). Chapter 2 below covers experiments with market power.
11. We are pleased to note that the model of random behavior of economic agents has a precedent. Leonid Hurwicz, who was our professor at Minnesota and was later awarded the Nobel prize in economic science, had conceived a process, which he called the B-process, where bargaining agents randomly choose bundles of goods. See Hurwicz, Radner and Reiter (1975a, b), where a stochastic process converges to outcomes that are Pareto efficient. Perhaps the coincidence was not due to chance, given that Shyam Sunder also taught at Minnesota.
12. The intuition for the result is probabilistic. It is likely that the first transactions involve seller robots with low valuation and buyer robots with high valuation. After some time, the seller robots that remain in the market are likely to have relatively low valuations, whereas the valuations of the buyer robots that remain are relatively high. Whatever transactions are realized at that point will take place at prices similar to the competitive equilibrium.
13. Indeed, the metaphor is even more powerful in our case, because Adam Smith assumed that, for the "invisible hand" to be effective, economic agents had to act with the aim of achieving maximum profit: "*By preferring the support of domestic to that of foreign industry, he intends only his own security; and by directing that industry in such a manner as its produce may be of the greatest value, he intends only his own gain, and he is in this, as in many other cases, led by an invisible hand to promote an end which was no part of his intention*" Adam Smith, 1776, Book IV, chapter II, paragraph IX. Now it

turns out that the invisible hand also seems to act even when but no one seeks his own gain! What a surprise!

14. Economics is the science of want and scarcity. Bosch-Domènech and Sunder (2000), generalizing the results of Gode and Sunder (1993a, b) show that want and scarcity, operating in a double auction market, are sufficient to attain the competitive equilibrium in an economy comprised of multiple inter-related markets. In short, even in complex economic systems, this type of equilibrium seems to be reached under assumptions on the behavior of the agents which are not particularly restrictive.

15. We have mentioned markets for goods, services and assets. Can we think of markets for something else? How about a marketplace of ideas, where ideas are exchanged? According to Matt Ridley (2010), it is precisely because of the exchange of goods, but also of ideas, among species members, that the human is the only species that has progressed throughout its existence.

16. For a simple experiment verifying the principle of comparative advantage, see Bergstrom and Miller (1999). This experiment is explained in detail in Chapter 8

17. The phrase "*one cannot understand theory if one cannot create it*" was written on the famous physicist Richard Feynman's blackboard at Caltech at the time of his death in 1988, see Stephen Hawking (2001, p. 83). It is indeed difficult to appreciate all the consequences of a theoretical model without developing its ramifications in the laboratory.

18. For experiments in macroeconomics, see Chapter 8 below. For finance see Chapters 4 and 5 above. For the economics of information, Chapter 9 below. And for applications to economic policy see Chapters 8 and 10 below and chapter 10 of Vol. 1.

2
Non-Competitive Markets
Nikolaos Georgantzís and Giuseppe Attanasi

Introduction

A market is not competitive when the agents acting in such a market have the power to influence the price, directly or indirectly, something that does not occur under perfect competition. Generally, these agents have market power because they are few in number, have access to relevant information and can foresee the interdependence between their strategies and those of others.

Among all the paradigms in economic theory, the theoretical predictions of oligopoly were the first to be examined in the laboratory. In the origins of experimental economics one can find the works of Chamberlin (1948) and of Smith (1962, 1964), who designed experiments to study a market with few agents that could reach the competitive equilibrium (see Chapters 1).

In this chapter, instead of surveying all experiments with few sellers,[1] we will adopt a narrower definition of the term *"oligopoly,"* and will focus on experiments that were directly inspired by the basic oligopolistic models of Cournot (1838), Bertrand (1883), Hotelling (1929), von Stackelberg (1934), and similar. We will omit, therefore, other experiments, such as those of Chamberlin and Smith, which were designed with the aim of testing the predictive power of the competitive equilibrium model.

Most of the experiments we consider in this chapter have been run in the last three decades.[2] This literature can be considered as a new wave of experimental work, aiming at representing basic oligopolistic markets and testing their properties. This work represents a systematic attempt to study a similar, but not identical, question to that tackled by Chamberlin (1948) and Smith (1962, 1964). While the latter compared the results in the laboratory with predictions of the *competitive equilibrium*, the series of experiments we review here compare observed behavior with the corresponding *oligopolistic equilibria*.

The chapter is divided into independent sections, which refer to different parts of oligopolistic theory: including monopoly and a number of extensions of the basic models that have been chosen with the aim of providing a representative overview of experimental findings in this area.

Monopoly, price competition and product differentiation

The simplest and most frequently corroborated hypothesis regarding price-setting behavior in markets argues that, by relaxing the usual assumptions, we may converge to the equilibrium price of Bertrand. That is:

- If there are (alternative assumptions): few agents, with no experience and limited cognitive ability, with insufficient information of the conditions of the market,
- is the equilibrium reached through learning by trial and error?

That is, whether the experimental subjects reach the price predicted by theoretical models that assume (usual assumptions) perfect information and infinite cognitive capacity.

The simplest case is that of *monopoly*. Assume that you are participating in an experiment in which you are a monopolist and you face a demand function unknown to you. In fact, this function could be, for instance, $Q = 100 - 2p$, which assumes that sales in your firm decrease when the price increases. Given that this demand function is unknown, you can only try to guess the price that maximizes your profits by testing what happens as you set different prices. For simplicity, let us assume that your cost is $C = 0$; thus, your payoff from the experiment coincides with your profits.

With an *unknown demand function*, you cannot calculate the optimal price, as you would do during a microeconomics class. As anticipated above, one way to calculate which price maximizes your profits as fast as possible is to set any price and observe the sales and the profits. Then you would change your price to see the reaction in sales and profits. Subsequently, you would use the results of these first two periods to choose the price for the third period, and so on. By accumulating information, you could start noticing how to get closer to the profit-maximizing price. Classroom experiments with college students have shown that the optimal price $p^* = 25$ is reached after six to ten trial periods.

Therefore, for a monopoly with no information on the demand function, laboratory experiments confirm that learning by trial and error leads to convergence on the perfect information prediction. This happens with few repetitions and without requiring specific cognitive abilities.

However, the result is different when some complexity in the problem facing the monopolist is introduced. For instance, suppose that, as a subject participating in the experiment, you are asked to simultaneously set the *prices of two products*.[3] To better understand the additional difficulty, consider a firm that aims to find the optimal prices for its two substitute products/services, such as two different flights to one destination operated by the same airline. In this framework, experiments have demonstrated a significant deviation in the strategies a subject plays with respect to the one predicted by the theory. The deviation lasts even after several attempts, unless the subject receives information on the cross effects of the demand of a product on that of the other product.[4]

In the real world monopolists frequently face decision-making problems that are even more complex, like when they have to set prices to manage a dynamic system: for instance, a renewable resource. García-Gallego *et al.* (2008) study a market of this sort, and find that subjects systematically fail to learn, and are unable to converge towards the optimal level of resource preservation, even when playing for 50 periods.

In summary, we can argue that human subjects can learn how to set the optimal price in a monopoly without the (extremely high) level of information and rationality assumed by the theoretical models in the textbooks. However, this is true only as long as the complexity of the environment does not exceed a certain limit. Above this limit, a trial-and-error learning procedure is not enough to allow experimental subjects to arrive at the monopolistic equilibrium price.

Let us now analyze the problem of price-setting in the case of an *oligopoly*. Fouraker and Siegel (1963) ran the first experiments on this topic. Their focus was on the importance of the type of information transmitted between subjects playing in the role of firms in an oligopolistic market. Their objective was to evaluate the predictive capacity of the theoretical oligopolistic equilibrium. Their experiments confirm two results of enormous interest:

i.   When subjects receive private information on their own profits, they tend to converge towards the Nash equilibrium prices (Bertrand-Nash equilibrium).
ii. If subjects also have available information on the profits of their competitors, then they will set prices higher than those predicted by the non-cooperative Nash equilibrium: their behavior shows a certain level of collusion.

The level of *collusion* dramatically increases if experimental subjects are given the possibility to communicate among themselves. Fonseca and Normann (2012) compare pricing behavior with and without the possibility of communicating between firms in Bertrand oligopolies with various numbers of firms. They find strong evidence that *communication* helps to obtain higher profits for any number of firms in the market: communication helps firms coordinate in collusive pricing schemes. However, the gain from communicating is non-monotonic in the number of firms, with medium-sized industries having the largest additional profit from communication. They also find that industries continue to collude successfully even after communication is prevented.

Let us now focus on the *demand side*: the most recent oligopoly experiments have adopted mechanisms of market clearing where the buyers' behavior is simulated with continuous demand functions.[5] For instance, García-Gallego (1998) uses a system of symmetric demand functions, composed by n equations as the following:

$$q_i = a - b \cdot p_i + \theta \cdot \sum\nolimits_{j \neq i} p_j$$

where p_i is the price of the variety i, n is the number of varieties available in the market, (each variety offered by a different firm), while j represents each of the substitutes of variety i.

The function tells us that the quantity demanded q_i of the good of variety i decreases as its price increases (b is a positive number), but it also increases as the price of any other variety increases. The parameter θ indicates the level of interdependency between varieties, in this case between variety i and the other $n - 1$ varieties. Such a demand system, where θ is positive and lower than b, corresponds to the case where each variety can be imperfectly substituted by any other variety in the market. Further, the unitary cost c is assumed to be constant and the same for all varieties, and there are no fixed costs.

In García-Gallego's experiment (1998), even though subjects systematically attempt to tacitly coordinate in setting prices at a level above that of the non-cooperative equilibrium, it is found that the Bertrand-Nash equilibrium strongly attracts individual strategies, especially in the second half of the market periods. This work shows that a 35-period horizon is sufficient to allow most subjects to converge surprisingly closely on Bertrand's prediction, with some experimental sessions in which predicted and observed behavior coincide. Figure 2.1 shows the evolution of prices in a typical session of the experiment.

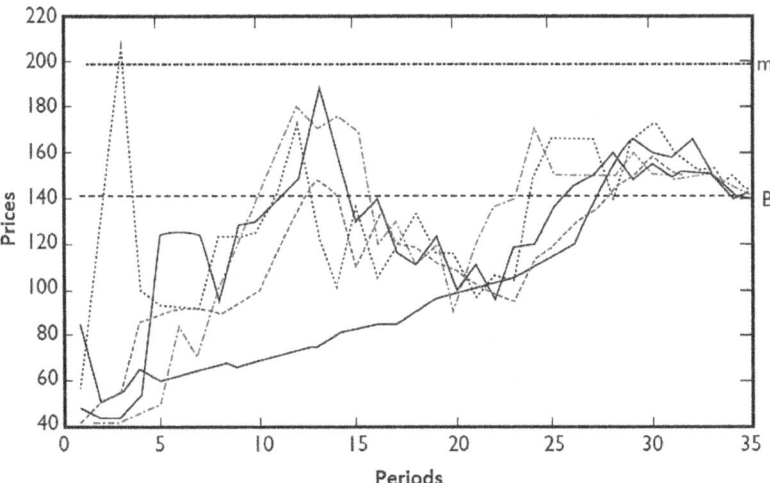

Figure 2.1 Bertrand-Nash vs. actual behavior in a differentiated oligopoly with five varieties

Note: After few periods of price volatility, subjects' strategies clearly converge to the Bertrand-Nash prediction (B) and move away from the monopoly (collusive) price (m).
Source: García-Gallego and Georgantzís (2001b)

Subsequently, García-Gallego and Georgantzís (2001a) implement the same conditions of demand and costs as those in García-Gallego (1998) to test the predictive power of Bertrand-Nash with *multiproduct firms*. The theory predicts that, if the same multiproduct firm jointly produces two or more substitute goods, their prices will be higher than if independent competitors offered the goods. Although the problem of a multiproduct firm is much more complex than that of a firm with a single product, the existence of a multiproduct firm leads to the prediction of an asymmetric Bertrand equilibrium, where the firms producing more goods tend to set higher prices. Surprisingly, the experiments in García-Gallego and Georgantzís (2001a) suggest that multiproduct firms do not understand the strategic profits they can derive from their multiproduct market power. This is why they behave as if their products were competing with each other.

Therefore, trial-and-error learning in an oligopolistic market with uniproduct firms leads to strategies that are close to the Bertrand-Nash equilibrium. However, in the multiproduct case, trial-and-error learning does not support the corresponding asymmetric Bertrand-Nash equilibrium. Davis and Wilson (2005) have reinterpreted this result in terms of its consequences for mergers policy: if firms do not realize their market power, behavior after the *merger* can be as competitive as before the merger.[6]

However, García-Gallego and Georgantzís (2001a) also carry out a treatment imposing an exogenous norm that limits the multiproduct firm to change its own prices all in the same direction (*price parallelism*). In this treatment, subjects tend to adopt the limit strategies of the Bertrand-Nash multiproduct equilibrium (that is, close to the collusive equilibrium)[7]. Figure 2.2 presents examples of sessions with multiproduct oligopolies.

The reader can identify that the prediction of the Bertrand-Nash multiproduct equilibrium for a firm that jointly sets the prices of three varieties (B3 in Figure 2.2) is the highest, followed by the same prediction for two varieties (B2), and that both of them are greater than the Bertrand-Nash equilibrium with uniproduct firms (B). In Figure 2.2 on the left hand side, a typical session is shown where, thanks to the exogenous imposition of the price-parallelism norm, firms' prices converge to the collusive equilibrium. On the right hand side, a session is shown where, without such exogenous imposition,

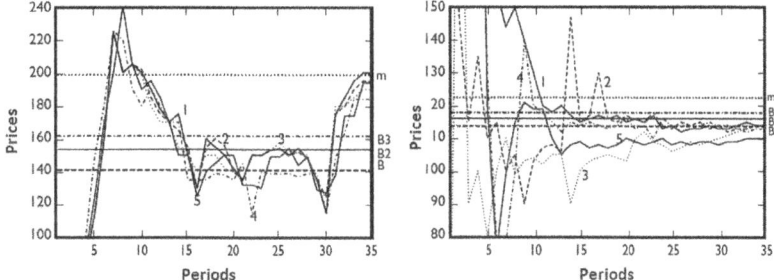

Figure 2.2 Multiproduct oligopoly

Note: Left: Convergence to the collusive equilibrium price (*m*) in a differentiated oligopoly with multiproduct firms, five varieties, and the price-parallelism norm. Right: Failure in reaching the collusive equilibrium price (*m*) because of no adoption of a price-parallelism norm.

Source: García-Gallego and Georgantzís (2001b)

multiproduct firms do not adopt price parallelism and cannot avoid converging to the Bertrand-Nash equilibrium with uniproduct firms.

Likewise, firms can affect the level of market competition by strategically choosing to differentiate one from the other. In the standard *product differentiation* model of Hotelling (1929), firms decide first on their *location*, which represents a variety in a continuous and closed product space, and then they compete in price-setting.

As in the case of several other phenomena for which it is very difficult to empirically test economic theories with real data, models with product differentiation have also been examined in the laboratory. Brown-Kruse and Schenk (2000), Collins and Sherstyuk (2000), and Huck *et al.* (2002b) have experimentally studied spatial markets with two, three and four firms, respectively.

These three works describe experiments in which the participants only choose the location of their firm, while the *prices are exogenously imposed* on them. In this context, two clear-cut theoretical predictions can be provided:

- There is a "minimal" product differentiation as in the non-cooperative equilibrium when subjects cannot communicate among them.
- There is an "intermediate" differentiation due to collusion when subjects can communicate between them.

Both theoretical predictions have been corroborated by the above-mentioned experimental studies. That is, firms that use location as the only strategic variable, in absence of communication tend to agglomerate in the middle of the product space (segment). Likewise, as we will see in Chapter 9, political parties tend to adopt ideological positions close to the median voter's preference. In the presence of communication, firms are located between the extremes and the middle of the segment.

However, the assumption of exogenous prices does not allow us to deal with the usual intuition: a firm can improve its profit by differentiating its product from that of its competitors with the intention of cooling down price competition. Recently, Barreda *et al.* (2011) have implemented experimental spatial markets with *endogenous price-setting*. This work presents two interesting results:

- There is a positive relation between differentiation and price.
- Differentiation by location tends to be low.

Figure 2.3 presents the aggregate results of the location – and price-setting stages, respectively. The results appear to be robust to variations in the experimental conditions regarding the rule of division of demand in case of a tie (automatized *vs.* human consumers).

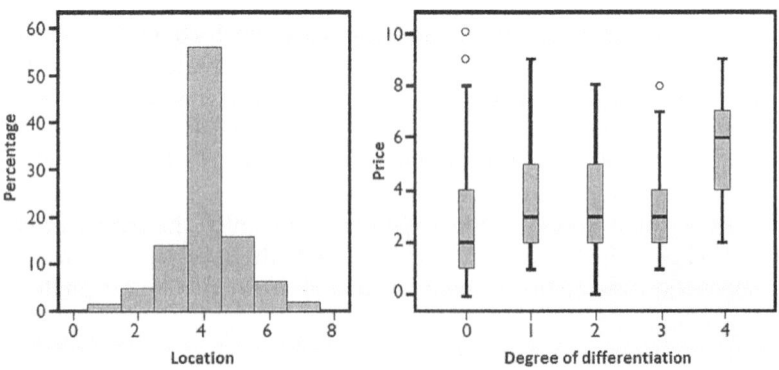

Figure 2.3   Spatial location with endogenous prices

Note: Left: The prediction of minimal differentiation by Hotelling (1929) is confirmed by the modal choice in an experiment with endogenous prices. Right: The hypothesis that prices increase for higher levels of differentiation is confirmed.

Source: Barreda *et al.* (2011)

In summary, the prediction of Hotelling's model, arguing that firms will agglomerate in the center of the market, is confirmed: both in experiments, where it is possible to compete in prices, as well as in contexts where the price is exogenously fixed. In addition, when price competition is allowed, subjects choose a product-differentiation strategy: they recognize its ability to sustain higher prices. It is very satisfying for a researcher to observe the predictive power of a theory in such a complex setting as the two-stage game – location and price – in Barreda *et al.* (2011), where the predictions of Hotelling (1929) are confirmed.

Finally, Camacho-Cuena *et al.* (2005) move one step further: spatial markets with endogenous prices, but in a context where the *location of the consumers is also endogenous*. Their experiments show that trial-and-error learning is not enough for consumers to understand that they should locate in the middle – between two sellers – which would increase competition and reduce prices.

In both experiments with endogenous prices mentioned here, coordination failures constitute an obstacle for sellers who attempt to avoid central locations to differentiate from other sellers.

Oligopolies with fixed quantities and extensions

The oldest oligopoly theory is the model of Cournot (1838), which differs from what has been surveyed up to this point due to the assumption that *firms compete by quantity* rather than price. The markets of Cournot (1838) have been experimentally studied by Fouraker and Siegel (1963), Holt (1995), Rassenti *et al.* (2000), and Huck *et al.* (2000, 2001b).

As discussed at the beginning of the chapter, a very general result of Cournot's experiments is that subjects, by adopting trial-and-error learning, show behavior that tends to confirm the theoretical predictions of Cournot-Nash. Although this is true on average, there is also certain variability around these predictions. These persistent oscillations decrease the predictive power of the equilibrium. In addition, in repeated-game scenarios, the total quantity is frequently not significantly different from the collusive prediction. In some cases, the total quantity oscillates between the collusive quantity and that of Cournot.

In summary, although moderately positive results regarding the predictive power of Cournot equilibria have been found, it seems that there are fundamental differences in the patterns of the data obtained in the experiments based on *Cournot models* (strategic quantity-setting) and the ones based on *Bertrand models* (strategic price-setting).

Huck *et al.* (2000) and Altavilla *et al.* (2006) both study markets of price and quantity-setting in a framework of differentiated products, where the Bertrand equilibrium is quite close to the Cournot equilibrium. Even though such experiments generally provide evidence that supports the Nash equilibrium predictions for the two types of markets (price-setting and choice of quantity), these works show that information derived from past strategies and results plays a crucial role.

Collusion in Cournot markets, *vs.* Bertrand markets, deserves thorough discussion. Suetens and Potters (2007) show that behavioral outcomes in Cournot markets tend to be more competitive relative to equilibrium as compared to those in Bertrand markets. Hence, more collusive behavior is detected on average in price-setting than in quantity-setting oligopolistic markets, with prices in price-setting experiments being above equilibrium prices, and quantities in quantity-setting experiments being above equilibrium quantities (see also Holt 1995, and Engel 2007). Moreover, the scope for tacit collusion in both types of markets is strongly affected by the number of competitors. Basically, implicit collusion is frequently observed in markets with two firms, rarely in markets with three, and almost never in markets with four or more firms. This effect has been observed under both Cournot competition (see, e.g., Huck *et al.*, 2004b) and Bertrand competition (see Abbink and Brandts, 2005, 2008; Orzen, 2008).

A famous modification of the Cournot model of quantity competition is due to Stackelberg (1934), which assumes that one of the two firms (the leader) chooses and commits in a credible way to produce a certain quantity, before the second firm (the follower) chooses its quantity. This model predicts that total production will be greater than that of the symmetric Cournot model and that, in general, the leader will produce and earn more than the follower. Huck *et al.* (2001a) experimentally compare the markets of *Cournot* and *Stackelberg*. Their results confirm the existence of volatility in the quantities chosen, similar to what is obtained in other experiments of production decision-making. In addition, the prediction that Stackelberg markets produce a higher total quantity is confirmed and that, therefore, such markets are more efficient than Cournot markets. However, the asymmetric nature of the interaction between the leader and the follower seems to hinder the convergence of the observed behavior towards the theoretical predictions of the Stackelberg model.[8]

Other experiments allow subjects to endogenously choose the moment they would make their strategic decision.[9] That is, before choosing quantities, the firms choose when to produce: in an early or a late period. In these experiments, simultaneous (Cournot) and sequential (Stackelberg) oligopolies can emerge endogenously. On the other hand, Huck *et al.* (2006) experimentally study a spatial market with *endogenous time choice*. In line with actual political campaigns, candidates can decide endogenously when and where to locate. Their results show that allowing for endogenous timing can eliminate some of the more unappealing equilibrium characteristics of the standard model.

In all these environments, the typical asymmetric results corresponding to the leader-follower structures receive less support than expected. In fact, the evidence is in favor of the symmetric results. This even occurs when the corresponding equilibria predict structures of the leader–follower type.

Cournot markets with *multiproduct firms* have received less attention in the experimental literature than corresponding Bertrand markets. A recent experiment by Hinloopen *et al.* (2014) analyzes the impact of product bundling in quantity-setting oligopolistic markets. One firm has monopoly power in a first market but competes with another firm à la Cournot in a second market. They compare treatments where the multi-product firm: always bundles, never bundles, and chooses whether to bundle or not. They also contrast simultaneous (Cournot) to sequential (Stackelberg) moves in the duopoly market. Their data support the theory of product bundling: with bundling and simultaneous moves, the multi-product firm offers the theoretically predicted number of units. In the case of sequential moves, when the multi-product firm is the Stackelberg leader, the predicted equilibrium is better attained with bundling, although this equilibrium is the same with and without bundling.

Oligopolistic models with *vertical relationships*, such as those between a manufacturer and a retailer, have received less attention by experimentalists than models with horizontal relationships. The basic model predicts that firms that are vertically related and not coordinated will set a total price–cost margin higher than if they agreed on a joint margin. This phenomenon is known as *double marginalization*. Durham (2000) experimentally confirms the importance of the double marginalization phenomenon predicted by the theory. Martin *et al.* (2001) corroborate these ideas in the lab and confirm the predictions of the theory of vertical integration as a means to exclude competitors. Furthermore, Normann (2011) shows that Bertrand markets with a vertically integrated firm are significantly less competitive than those where firms are separate. However, he shows that, while his experimental results violate the standard Nash-equilibrium notion, they are consistent with the quantal-response generalization of Nash equilibrium.

Other interesting extensions of the basic oligopolistic model are concerned with the *value of delegating* firm's strategic decisions to managers whose incentives are designed to either pursue or deviate from the firm's profit maximization. For instance, rewarding a manager according to the firm's production, rather than according to its profits, makes the firm a Stackelberg leader. Huck *et al.* (2004a) test the influential theory of Vickers (1985) and that of Fershtman and Judd (1987) on the strategic role of delegation in oligopoly through the design of incentives for the managers. These experiments study a situation in which the owners of the firm choose whether their managers receive as compensation either a share of the firm's profits or a reward based on the firm's revenues. Surprisingly, the second option is chosen very few (5%) times.

Georgantzís *et al.* (2008) study an *endogenous compensation system* in an experiment where the firm owners can choose between offering their managers compensation dependent on a linear combination of their own firm's profits and revenues, or compensation dependent on their own firm's profits relative to the profits of rival firms. As the theory presented in the paper predicts, the preferred remuneration system is the one based on own-firm profits, compared to those of rival firms.

Other (ir)regularities

A recurrent aspect in all oligopoly experiments, although it does not constitute a central topic in most of them, is *learning*. In fact, the trial-and-error learning process determines, to a great extent, which outcome is obtained in an oligopolistic market.

Cyert and DeGroot (1973) made an important contribution in this direction, by relating learning to the ability duopolists have in reaching a collusive outcome. It may be that learning in this context does not imply a disclosure of the mathematical properties of the supply and demand model the firm faces. In simple terms: learning is a dynamic process in an individual's decision-making. It is a process that leads him/her from initially uninformed strategies towards the relevant region of cooperative or non-cooperative equilibrium.

Multiple studies have attempted to identify possible systematic patterns in the learning strategies people have. Many researchers have analyzed adaptive learning (see, e.g., Nagel and Vriend 1999). A common result is that it does not seem that people learn through sophisticated or formal processes. For instance, García-Gallego (1998) and García-Gallego and Georgantzís (2001a) offered participants the opportunity of obtaining linear estimates (from an ordinary least squares model) of the underlying demand and, also, gave them various graphical representations (quantity-price, profit-price, etc.) of the data from the previous market period. The conclusion in both studies is clear: subjects did not make any effort to systematically calculate the optimal strategy by using explicit optimization, despite the fact that participants were academically advanced students (some of them were even graduate students in economics).

However, learning significantly affects observed behavior, as in most experiments participants' strategies first show a high degree of dispersion (they look almost random), and evolve over time towards the reference solutions, such as the collusive or non-cooperative outcomes. This brings us to an important determinant of the collusive outcome in oligopoly experiments. Mason and Phillips (1997) confirm the importance of information in a duopoly with asymmetric costs. In general, it is important to distinguish between *two sources of information* provided to participants in an experiment:

- Information can be provided *ex-ante*, through the instructions (at the beginning of the experiment).
- Information can be provided *ex-post*: that is, it becomes available as a result of past choices (feedback given during the experiment).

Contrary to what has been suggested by theorists, experimental treatments where information is given prior to the experiment have very little or no effect at all upon observed behavior: subjects hardly ever use information that is not immediately interpretable in their decision-making process. Therefore, information about the exact conditions of supply and demand, for instance, has very little effect on observed behavior. This is especially true when these conditions do not provide a "linear" interpretation to supply and demand.

On the contrary, informing participants about the strategies chosen, and even more importantly, about the outcomes obtained by their competitors in previous periods, has a significant effect.

Knowing the outcome of previous choices is not the only source of learning in oligopoly experiments. Various experiments have identified learning processes different from those assumed in the theoretical models. For instance, Huck *et al.* (1999) show that information about other players' strategies plays an important role in the emergence of collusive (rather than non-cooperative) outcomes. In addition, imitation of the most successful competitors appears to be supported by some experimental evidence.[10]

Another important aspect, that systematically affects behavior in oligopoly experiments, is *inequity aversion* (see chapter 6 of Vol. 1). Inequity-averse subjects tend to pursue payoffs similar to those of the competitors, even when the experimental context is initially asymmetric. Admitting a certain level of participants' inequity aversion can explain all those cases in which the theory fails to predict the observed behavior in asymmetric oligopoly experiments. The result of Huck *et al.* (2001b) explicitly relates inequality in payoffs with lack of stability in the Cournot setting. Altavilla *et al.* (2006) find that informing the oligopolists about past prices set by their competitors leads to quantities closer to those predicted by the Cournot-Nash equilibrium, while providing information about the average profit of the entire industry leads to higher levels of cooperation.

Aspiration levels of oligopolists are also constitute an important factor when looking at the divergence between theoretical and experimental outcomes. Huck *et al.* (2007) show that aspiration levels can be used to explain the merging paradox – where the merged firm ends up earning less despite there is less competition in the market after the merger – observed in the laboratory. Indeed, the aspiration-level hypothesis predicts that after the merger a firm has a target profit in mind, the one obtained before the merger, and it will also act in order to maintain it after the merger.

Finally, the role that *risk aversion* has on strategic behavior is evident in Sabater-Grande and Georgantzís (2002), showing how more risk-averse individuals have a lower probability of cooperating in a Prisoner's Dilemma (see chapter 7 of Vol. 1). The latter can be interpreted as a limit case (with only two strategies per player) of a Bertrand or a Cournot standard duopoly. The management literature is full of business cases identifying over-risky decisions made by managers. If we acknowledge that the managers may deviate from the pure profit maximization due to their aspiration levels, for personal and psychological reasons, or due to the incentives in their management contracts, the idiosyncratic effects observed in oligopoly experiments can be especially relevant to decisions that firms make in the real world.

The issue of whether the results of laboratory experiments on oligopolistic markets also extend to comparable (real) situations *outside the laboratory* is discussed in Potters and Suetens (2013). They correctly emphasize that there are several dimensions to this concern. In particular, decision-makers in firms are not students, and firms' decisions are usually not made by one individual acting on his/her own behalf. As to the latter concern, some experimental studies have begun to explore decisions by groups of individuals (boards) and how these depend on the decision-making process in the group. As Potters and Suetens (2013) report, in some cases individuals seem to act very differently from groups, whereas in other cases few differences are found.

Conclusions

Experiments on oligopolistic markets have aimed at testing the predictive power of oligopoly theory in explaining observed behavior in experimental settings that implement the conditions established by each model. Although such aims may appear to have limited relevance to the world outside the laboratory, this line of research has taught us some very interesting behavioral principles. For example, the oligopolistic equilibrium may be the limit towards which the strategies of economic agents who learn by trial and error converge. Further, such a learning process has a higher probability of supporting symmetric than asymmetric theoretical predictions. Finally, this learning process, as well as some idiosyncratic features of the experimental participants, can either help or hinder the corroboration of the theoretical predictions. Throughout this chapter we have tried to make it clear that we remain closer to the beginning than the end of this thrilling process of understanding how imperfectly competitive markets work.

Notes

1. According to the Greek etymology of the word, oligopoly (ολιγοπωλιο) is a market with few sellers (ολιγοιπωλητές).
2. The first experimental tests of Industrial Organization Theory can be found in Plott (1982). Later, Holt (1995) presented a summary of the experimental results in oligopolistic markets.
3. Kelly (1995) is one of the first examples of a multi-product monopoly experiment.
4. See for instance García-Gallego *et al.* (2004).
5. In other experiments subjects are provided with discrete payoff matrixes, which are the reduced version of the original oligopoly games with continuous strategies. The choices for a more or less realistic experimental design depend on the principles followed by the experimentalist and on his/her research objectives.
6. Fonseca and Normann (2008) provide a thorough analysis of the impact of *mergers* in experimental Bertrand oligopolies. They consider as treatment variables the number of firms (two, three) and the distribution of industry capacity (symmetric, asymmetric). They find that, even though they are more concentrated, asymmetric markets exhibit lower prices than symmetric markets with the same number of firms. Consistent with the static Nash-equilibrium prediction, duopolies charge higher prices than triopolies. However, although the overall impact of a merger is anti-competitive, the price increase is not significant. This last result, in a sense, confirms the findings of Davis and Wilson (2005).
7. Price parallelism in uniproduct firms has been previously studied by Harstad *et al.* (1998) in a context specifically designed to tackle this question. It was found that the conscious adoption of price parallelism by the competitive sellers had the effect of increasing prices towards the collusive prediction.
8. Kübler and Müller (2002) analyze experimentally markets with price-setting designed with the aim of comparing simultaneous and sequential decisions. They highlight the difference between authentic sequential games and sequential strategies obtained through the "strategy method" (subjects are asked what they would do for each choice profile of the other subjects and then the binding case is randomly chosen).
9. Huck *et al.* (2002a), Fonseca *et al.* (2005, 2006), and Muller (2006) are some of the experimental studies; see also Normann (2002) for a theoretical analysis.
10. See Offerman and Sonnemans (1998), Offerman *et al.* (2002). For a contrary option see Bosch-Domènech and Vriend (2003).

3
Economic Systems Design

Diego Aycinena, David Porter and Stephen J. Rassenti

Introduction

In 1992 the Cassini mission destined for Saturn was preparing to transport dozens of scientific exploration instruments with a constrained budget and limited resources. Each team of engineers and scientists was competing for resources (mass, energy, data transmission, etc.) against the other teams. When the Jet Propulsion Laboratory from NASA contacted a team of economists to help them design a mechanism that would ease the allocation, one of the economists immediately replied: "Let them trade." Facing the trivially correct answer, the engineers insisted: "Could you be a little more specific?"

The engineers were not expecting a correct but abstract principle about how to solve a general problem, but a specific allocation or exchange system that would fulfill certain objectives for the concrete problem they were facing (see Wessen and Porter 1998, 2007). The answer should then be drawn from the field of Economic Systems Design (ESD from now on).

ESD is a new economic application that seeks to design mechanisms to solve resource allocation problems. The application of ESD comes from mechanism design theory[1] and auction theory,[2] which are complemented with economic experiments. Both areas aim to analyze resource allocation problems when information is dispersed between the agents: that is, when it is necessary to require private information from different agents to make an optimal allocation, agents who may have incentives to not truthfully reveal their information.

ESD is supported by economic experiment as a fundamental tool to complement theory and test the proposed mechanisms at laboratory scale. ESD has emerged as an answer to the demand for new methods of organizing markets, for creating new exchange mechanisms and new forms of buying and selling goods, or selling or buying rights or services, that have never been exchanged in markets before. Some examples are: pollution emission permits; the generation, transmission and distribution of energy; the auction of rights to use the electromagnetic spectrum, and similar topics related to telecommunications; oil pipeline networks; rights for the use of runways in airports; and recently, the new services of electronic commerce and (positions of) announcements in search engines on the Internet.[3]

Conceptual framework

Before analyzing ESD and its relation to experimental economics, it is advantageous to know some terminology and its conceptual framework. Smith (1982, 1989) provides the conceptual framework for laboratory experiments, based on the elements needed to define a microeconomic system:

- the environment,
- the institution, and
- agents' behavior.

The *environment* defines the agents that participate, with their respective values, costs, information, technology and resources. A monopoly or a duopoly, bidding for homogeneous or heterogeneous goods, with equal or different cost curves if the buyers value the goods differently, or if all of them value it equally, etc.

All these examples describe different environments. In the laboratory we can control the environment through monetary rewards to induce the value/cost structures we want for that laboratory economy (Smith 1976, 1982).

The *institution* or mechanism refers to (the algorithm defining) the rules and procedures under which the agents interact. That is, the rules by which agents communicate, exchange and/or produce to modify their initial allocation. We may speak both about rules regarding the conditions and the types of messages, as well as the procedures to process these messages and convert them into trans-actions, allocate goods and resources, impute prices and costs, and manage information. Consider the following examples of different auction formats:

- Open outcry *English* auction: the participants can send, as a message, any bid price that is higher than the last bid made, and this goes on until the auction ends.
- *Dutch* auction: the asking price starts at an arbitrarily high level and starts descending until a first participant sends the message that he accepts a bid to purchase at that price, after which the auction ends.
- Sealed bid auction: each participant writes his bid (message) and tenders it in an envelope. All the participants make their bids simultaneously.

The allocation rule indicates how the good auctioned is allocated given the messages received: to the person with the highest bid, in the case of the English or sealed bid auction, or to the first person accepting a price, in the Dutch action.

The rule for price imputation establishes which prices each partici-pant has to pay given the messages that were sent. In sealed bid auctions, the auctioneer allocates the good to the participant who made the highest bid, paying their own bid – in the case of the first-price auction (FPA); or paying the second highest bid – in the case of the second-price auction (SPA). That is, in the SPA the winner of the auction is the person who makes the highest bid, but pays the price of the second highest bid.

Finally, the *behavior* refers to the strategic actions chosen by the (rational) agents, given the environment and the institution. These actions are revealed by the messages sent: bids, asks, acceptances, etc. The outcomes of a microeconomic system – such as, final alloca-tion, prices, costs, producer and consumer surplus, efficiency of the system, etc. – depend on how the rules of the institution process the messages sent by the agents in a specific environment.

Based on this conceptual framework, we can say that ESD is an *engineering* process with the aim of designing an institution or mech-anism that – operating in a particular environment – gives incentives to the agents to behave in a way that leads to the desired outcomes.[4]

What are laboratory experiments good for? From an initial design we can test the proposed mechanisms until we reach the desired results: maximizing efficiency, maximizing income/minimizing costs, reducing price volatility, maximizing the consumer surplus, etc. In the same way that aeronautic engineers use wind tunnels to test new aircraft designs, or engineers use load tests to verify the solidity of recently built bridges, laboratory experiments serve to test – in a controlled environment – different allocation mechanisms, based on models of the environment and with subjects motivated by real incentives.

The challenge of ESD consists, therefore, of designing mechanisms that can incentivize agents to truthfully reveal their private information, or designing an institution capable of reaching the desired outcomes even when the private incentives agents have may be to distort the information revealed by their actions.

Consider the case of a first-price auction institution in an environment where multiple buyers with independent private values[5] bid for a single good. Let us look at the incentives generated by the allocation rule – he who makes the highest bid wins – and the price imputation rule – the winner pays their bid. In an FPA:

- The allocation rule induces high biding: The bidders have incentives to increase their bids in order to increase their probability of winning.
- The price imputation rule induces low biding: The participants have incentives to reduce their bids because paying a high price, if they win the auction, reduces their payoffs. That is, the subject may obtain the good but if he has paid a very high price he will not win anything.

Consider the following simple environment. Some risk-neutral homogeneous actors whose valuation of the auctioned good follows a uniform distribution with values between 0 and $\bar{v}$. Each participant knows his own valuation of the good, but only knows the distribution of values for the other participants.

The bid (b) that a subject would make in a risk neutral Nash equilibrium (RNNE) will be a function of his value (v) and the number of bidders (n):

$$b(v) = \frac{(n-1)}{n}v.$$

In this case, we see that the subjects under-reveal their individual valuation of the good – given that $(n-1)/n$ is always lower than one – but such a valuation increases concavely in n.[6] That is, as the number of participants in the auction increases the bid will be closer to the real valuation.

What would happen if we changed the price imputation rule? Are we able to incentivize the truthful revelation of the private value with some variation of it? In the second-price auction (SPA) – also known as a Vickrey auction[7] – the price paid by the winner does not depend on his bid, but on the second highest bid. How does the optimal strategy change in this case? The weakly dominant strategy is to bid *exactly* the value of the good, regardless of the number of participants in the auction or their risk attitude. Therefore,

$$b(v) = v.$$

There is no doubt that this is the weakly dominant strategy for an SPA. However, multiple laboratory experiments have reported that it is difficult for participants to discover and follow such a strategy (Kagel *et al.*, 1987, Kagel and Levin 1993).

Multiple experiments have found mechanisms that work very well in the laboratory. Consider the case of an English auction. The predicted behavior in this auction is theoretically equivalent to that of the SPA: for both cases the theory predicts that the bids truthfully reveal the private value. Different experiments analyzing these auctions (Coppinger *et al.*, 1980, Kagel *et al.*, 1987, Harstad 2000) have found that the theory correctly predicts behavior in such auctions, and the participants' behavior quickly converges to the dominant strategy.

These examples of single good auctions provide useful illustrations of different aspects.

- First, we see how the incentives that generate the mechanisms affecting behavior, in particular regarding the information revealed by participants through their bids or messages.
- Second, we see how such behavior can generate different results (different efficiency levels, lower incomes for the auctioneer, etc.).

- Finally, it helps us to understand the fundamental role of experiments in ESD, supported by theory, in finding the mechanisms that generate the desired behavior and outcomes.

Given the nature of ESD and its emphasis on practical applications (as in the case of the Cassini mission), occasionally these types of experiments do not seek for new theoretical developments, but aim to make a comparative analysis of the performance different institutions have and/or of their robustness in different environments.

As the scope of ESD is very wide, due to space constraints we must leave out several interesting topics, such as the vast literature on single unit auctions and multiple homogeneous units auctions (Kagel 1995, Kagel and Levin 2008),[8] *matching* problems (Roth 2002), or work on information markets (Wolfers and Zitzewitz 2004, Tziralis and Tatsiopoulos 2007).

Thus, in what follows we will focus on experiments for ESD applications in complex settings and with different objectives: the efficient resource allocation of multiple goods with complementarities or substitutes through a combinatorial auction, and market allocation mechanisms of rights for generating electric power, seeking to mitigate the agents' inherent market power in this environment.

Combinatorial auctions

A combinatorial auction is an auction designed to allocate resources, which allow for the incorporation of logical constraints ("AND"/"OR") on the bids, so that an agent can make bids for packages of goods. This is extremely useful for the allocation of multiple resources that are complements or substitutes to each other.

The following example will help as an illustration. Assume an environment where three goods are auctioned (*A, B,* and *C*) and there are three agents who want to acquire them: Vicky, Cristina and Juan. Table 3.1 shows the valuations each individual has of the goods and on each of the possible combinations.

The table shows that in some cases there are complementarities (super-additive demand) and in some there are substitutes (sub-additive demand). Think, for instance, of Cristina's case:

Table 3.1   Complementarity and substitutability between goods

	A	B	C	AB	AC	BC	ABC
Vicky	100	50	20	290	230	170	290
Cristina	10	150	30	210	120	160	260
Juan	70	80	10	170	110	100	350

- Goods *A* and *B* have a super-additive demand, as their joint value is greater than the sum of the individual values: $AB = 210 > A = 10 + B = 150$.
- On the other hand, goods B and C have a sub-additive demand, as their joint value is lower than the sum of their separate values: $BC = 160 < B = 150 + C = 30$.

The table allows us easily to observe the utility of allowing logical restrictions on bids: that is, of making bids for *packages*. If it was not possible to bid for packages, but instead bids had to be made simultaneously for each individual good, how much would Juan be willing to bid for *A*, for *B*, and for *C*? The valuation of the three joint goods for him is 350. If he bids 100 for each and obtains all three (*ABC*), he would have a profit of 50 ($350 - 3 \times 100$); but if he only gets *A* and *B*, he would realize a loss of 30, as *AB* is worth only 170 ($170 - 2 \times 100$). And if he was to get only *C*, he would lose 90, given that *C* is worth only ten and he would end up paying 100.

This is known as the *exposure problem*: the agents are exposed to the risk of winning some, but not all the goods of a complementary set. A combinatorial auction eliminates this problem by allowing Juan to make a bid for the *ABC* package (*A* and *B* and *C*, *if and only if* he gets all three goods).

Similarly, we observe that Cristina's demand for the *BC* package is sub-additive. The exposure problem is also present here, because if she had to simultaneously make individual bids she would be exposed to the risk of winning both goods and incurring losses: for instance, if she bids 140 for *B* and 25 for *C* and she obtained both goods. This risk is eliminated in the same way by allowing for logical constraints on the bids: by doing so she could bid 140 for *B or* 25 for *C* (but – with those prices – she would at most want to get one of the goods).[9]

The example described in the previous table may seem to be an economist's illusion, but there are multiple valuable resources with similar characteristics. An illustrative example is that of licenses for using the radio electric spectrum for telecommunications. Such licenses are usually allocated to concrete regions that allow the use of different frequency bandwidths, creating complementary packages – adjacent regions of one same market – and substitutive packages – different frequency bandwidths for the same regions.

Another example is that of the rights for time slots to use airport runways for landing and takeoff. For instance, at La Guardia airport in New York, a time slot is a 15-minute interval, for which the maximum number of airplanes – which can vary given the weather conditions – is authorized to use the airport runway to land or take off. An airline requires (at least) two time slots in compatible schedules to be able to cover one route. Such time slot sets have a high degree of complementarity: they are worth much more together than the sum of their parts.[10] For instance, the right to use a runway for a time slot that JFK airport in New York together with that of a time slot (7.5 hours later) in Barajas, Madrid, has a high value for an airline, but one without the other has practically no value.[11]

Beyond real cases in environments with similar characteristics, multiple practical experiences witness the benefits of using combinatorial auctions. In Chile they have been successfully used in the allocation of contracts for the provision of school lunches to children of low-income families (Epstein *et al.*, 2002); in London to allocate bus routes for urban public transportation (Cantillon and Pesendorfer 2006); in the United States to award transportation services and logistics to companies such as Sears Logistics, Home Depot, Walmart, Compaq, Kmart, etc. (Ledyard *et al.*, 2002); and in Guatemala to allocate contracts to build and operate electric energy transportation lines (Argueta *et al.*, 2010).

Consider again the example in Table 3.1. To obtain a maximally efficient allocation – by allocating resources to those who value them the most – it is necessary to evaluate the different possible combinations.[12] After evaluating them, we observe that the optimal allocation is as follows:

- Vicky gets *A* and *C*,
- Cristina *B*,
- Juan gets nothing.

This allocation generates a total value equal to 380 (= 230 [AC for Vicky] + 150 [B for Cristina]). This is greater than the value of any other possible allocation. The real problem, nonetheless, arises because the information displayed in the table – the valuation each agent has for the different packages – is private and dispersed between them. Thus, to optimally allocate these goods we find ourselves again in need of an allocation mechanism able to incentivize agents' behavior that generate the outcomes we aim for. Porter, Rassenti, Roopnarine and Smith (2003, from now on PRRS), based on the results from previous experiments,[13] designed a *combinatorial clock auction* (CC).[14] In an ascending clock auction (or English clock), the price starts at an arbitrarily low price and increases as long as there is an excess of demand, that is, as long as the number of units demanded by the buyers at such a price is higher than the number of auctioned units. The participants may only leave the auction, indicating – as the price increases – that they do not want to buy at the current prices. Recall what was discussed in Chapter 1: some institutions can replace part of the intelligence required from the agents to find an equilibrium. Using a clock that increases the prices automatically with the excess of demand is an example of this type of institution.

In the case of a *CC* auction, it starts with a *price clock* for each good. All price clocks starts at arbitrarily low levels.

- In the first round, the participants indicate the goods they wish to buy; they are allowed to bid for packages, using logical restrictions such as (*A* and *B*) or (*B* and *C*), at the initial prices.
- In the following round, the *price clock* sets an increment in all those goods for which there is an excess of demand.

In cases where the demand for a good is exactly the same as its supply, prices stay the same as in the previous round. With these new prices the bidders indicate whether they are still interested in the goods or packages of goods, or if they wish to eliminate their bid for them. The subsequent rounds continue in the same way: the price for the goods with excess demand increases and the participants can leave the auction, which ends when the bid for each good eliminates the excess of demand.[15]

To test the correct functioning of this mechanism, PRRS designed laboratory experiments with a complex environment: ten heterogeneous goods were auctioned between ten participants with diverse demands for each good and/or packages of goods. They used seven variations of the environment, modifying it so that they could establish whether the CC auction was robust to these variations. They conducted the experiments with different environments, comparing the efficiency results generated by the CC with those of two other mechanisms of combinatorial auctions: the mechanism applied by the FFC to award the spectrum (SMR)[16] and an alternative mechanism proposed for such an award (CRA).[17]

The objective of PRRS was to maximize efficiency levels and, if we focus only on the results, it appears that they achieved this. The efficiency levels reached were consistently higher with the CC and clearly superior to those of the CRA and SMR. In addition to the high efficiency levels obtained, the CC auctions provided multiple advantages compared to the others:

- *Intuitive*: The process is simple and the cognitive costs for the participants are low: they only need to understand their own valuations of the different packages and to respond according to the relative prices they observe.
- *Feedback*: The prices the participants observe provide enough feedback to allow them to refine their strategies. On the other hand, feedback is limited (it does not provide information on the behavior of the rest of the participants) and it does not facilitate anti-competitive strategic behavior.
- *Limited field of messages*: If the possibility of information transmission is restrained, anti-competitive strategic behavior is also reduced: tacit collusion by sending signals and retaliations.

Despite the results obtained by PRRS, Ausubel *et al.* (2006) believe that the CC auction can be threatened by tacit collusion from the participants – withdrawing early from the bid for marginal units;[18] thus, they suggest a hybrid auction mechanism: a clock process, followed by a bid through a *proxy* for each package they are interested in.

It is still open to empirical verification up to what point the CC auction can be the object of tacit collusion and to what extent the mechanism proposed by Ausubel *et al.* (2006) can mitigate it. Laboratory experiments will surely be a fundamental tool for evaluating these still unanswered questions.

ESD and electric energy markets

We will now look at some cases of how ESD has been applied in electric power markets. The agents in the electric power environment are generators, distribution companies, transmission companies and consumers.

- The generators convert actual or potential energy into electricity, which they offer for sale in organized markets.
- The distributors attend to the demand of the regulated consumers – their obligation is to serve them.

The distributors are commonly price-takers, *passively* purchasing the amount of energy their clients consume, who usually have the right to consume almost any amount they want to – normally at prices that do not vary for weeks or even months.

There are some particular factors in this setting that draw attention to the market power of the generators:

i. The demand is very inelastic: it has little reaction to variation in prices.
ii. Often, there is a relatively low number of generating firms that control a high percentage of the generation capacity.
iii. Generators participate in a repeated game – which tends to facilitate collusion.
iv. Transmission capacity may be constrained, generating local market power in certain zones.

Intense experimentation effort has been concentrated on searching for mechanisms that allow us to mitigate the problems of market power. Among the main proposals that have been experimentally evaluated for controlling the market power of generating firms are: variations in the number of participants, the introduction of *forward* markets (LeCoq and Orzen 2006, Brandts *et al.* 2008, van Koten and Ortmann 2010, Ferreira *et al.* 2009),[19] and the active participation of the distributors in the market (Rassenti *et al.* 2003a).

In electric power markets, short-term demand tends to be highly inelastic because an important share of the demand is for regulated consumers, facing prices that do not reflect the time variations in the marginal costs of generation.[20] Rassenti, Smith and Wilson (2003) (RSW) used a 2 × 2 design, with two levels in the environment (with and without market power) and two levels in the role of demand (active and passive). The aim was to explore environments with inelastic demand and to compare the effects of allowing active participation in that demand – with the potential of making it more elastic.

In many cases the distributors' demand is passive, as they are price and quantity takers. The amount they "demand" is determined by the quantity consumed by their clients, and the price is fixed according to the marginal bid of the generators that satisfies such a quantity. In treatments with passive demand, they used robots that truthfully reveal the demand. As we saw in Chapter 2, this is a common procedure in experiments where the distributors are price-takers. In cases with active demand, the distributors would have contracts with some clients to whom they could interrupt (cut) the service in exchange for some compensation.[21] In treatments with active demand, they used subjects (motivated by payoffs) who controlled a small portion of the demand they had the option to "interrupt" at a cost. That is, in this case the distributors can modify the quantity consumed by their clients, and in that way attempt to influence the price.

To implement the variants of market power without modifying the aggregate supply, the authors reallocate, in a treatment, generation units with intermediate costs from sellers 4 and 5 to sellers 1 and 2 (as indicated by Figure 3.1). In this way the aggregate supply is kept constant, but manages to grant market power to sellers 1 and 2.

Observe Figure 3.1 in a little more detail. The competitive price is equal to the marginal cost of the generators and, in the periods of intermediate demand, this cost will be 76. The maximal price that guarantees 100% efficiency will be the (maximum) price that satisfies the entire demand. For such a period it would be 96 (the price of the last step of the intermediate demand). Observe that when sellers S1 and S2 have market power they can unilaterally withdraw their offers for the four intermediate units (their cost is 76). This would push the prices to the next step in the supply curve (price=166). Even though they would not sell any unit with the intermediate cost, each would sell the four units with a base cost of 20 at a price of at least 166 and, obviously, this would be a very profitable strategy: any of the generators can unilaterally increase their payoffs considerably by simply withdrawing their units, and offering them at a higher price.

The reader can do the following exercise.

- Determine that if sold at the competitive price of 76, the earnings for S1 or S2 would be: $(76 - 20) \times 4 = 224$.
- If you take away the intermediate units, they manage to raise the price to (the following supply step) 166, and their earnings would be: $(166 - 20) \times 4 = 584$.
- In the case where they sell all their units (the four base and the four intermediate) at the maximal price of 100% efficiency (in the third step of the intermediate demand) equal to 96, their earning would be: $(96 - 20) \times 4 + (96 - 76) \times 4 = 384$.

From which it should be clear what the best strategy is.

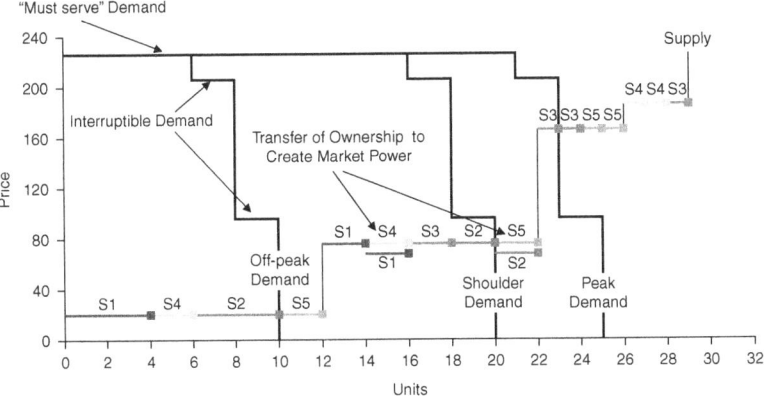

Figure 3.1   RSW experimental setting

Note: Figure 3.1 illustrates the three demand scenarios (base, intermediate and peak) and the aggregate supply in the experiments of Rassenti, Smith and Wilson (2003). In addition, it shows variations in the environment to apply treatments with and without market power: In the middle of the experimental sessions (treatments with market power), sellers 1 and 2 control four units with a cost of 76 each (in the intermediate demand scenario). In the other sessions (without market power), sellers 1 and 2 control only two units at a cost of 76 each; sellers 4 and 5 control the other two units. Furthermore, in all cases, seller 3 controls the other two units at a cost of 76.

The environment used by RSW is highly complex – each experimental session lasts 14 "days," and each "day" consists of a cycle with four periods of demand (intermediate 1, base, peak, intermediate 2).[22] Due to this complexity, they used doubled experimental sessions, one for training (with six bidders), and the second – two days later – to get more data (with the five best bidders in the previous session). The results of the experiment show that, in the treatment without market power and with passive demand, the results converge to the efficient price range – between the competitive price and the maximal price for 100% efficiency – in all periods of demand. When turning to the treatment with market power (where the generation actives are real-located), prices considerably increase in the periods of intermediate and base demand.

In treatments that facilitate active demand participation, for both cases (with and without market power) the results indicate that prices are consistently maintained in the efficient price range.

Figure 3.2 shows the prices for the four sessions of the treatments with market power, with and without active demand participation.

As can be observed in this figure, active participation of a fraction of the demand is enough to neutralize the bidders' market power. The

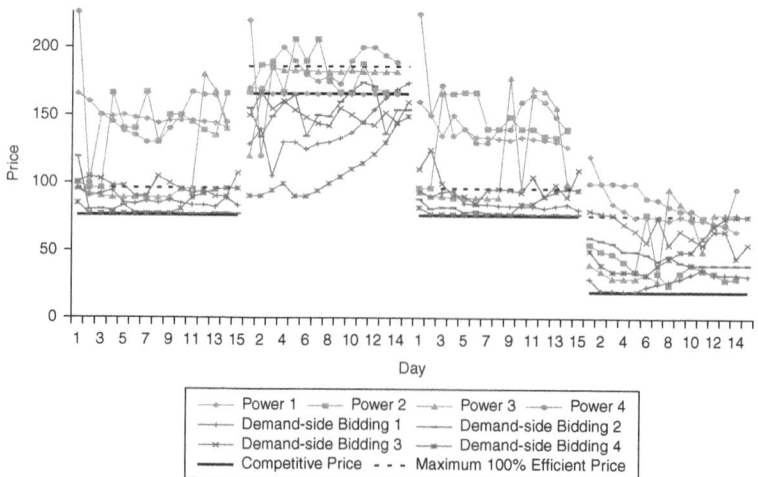

Figure 3.2 RSW treatment with market power, with and without active demand

results from RSW are consistent with those from other experiments (Denton *et al.*, 2001), and suggest that the design of a market where active participation in the demand is facilitated is a promising path to mitigate the market power of the supply.

In both cases, although the mechanisms were seeking different objectives, we have seen that the experiments play an important role. Through experiments we can compare different institutions while maintaining a controlled environment, or we can see to what extent a certain mechanism is robust to variations in the environment. In addition, they allow us to run controlled tests, to scale, before implementing a new design mechanism.

Figure 3.2 shows the prices (in the treatments with market power) in the four demand cycles under the three scenarios, in the RSW experiments. As indicated in the graph, prices in (both cycles of) the intermediate demand periods, with passive demand and market power, tend to be above the limits of prices with 100% efficiency. By allowing active demand participation it is possible to control the bidders' market power, reducing prices down to ranges of 100% efficiency.

Conclusions

In this chapter we have seen that experiments are a fundamental tool for ESD. The focus in this chapter has been different from that in other chapters, because here we have seen a more applied focus for experiments. After introducing ESD and its conceptual framework, we saw how experiments vindicated the design of a combinatorial auction mechanism for the efficient allocation of resources with complementarities. Subsequently we have seen the role that some experiments have played in ESD to evaluate a mechanism of mitigation over the power the sellers may have in the electric power markets. Due to space constraints, we have had to omit many interesting applications, but there is no doubt that it is becoming more common to use experiments to test novel designs that aim to solve practical problems. Such problems can range from the relatively mundane, such as allocating course slots in graduate programs in high demand (Denton *et al.*, 2001), to real life and death matters, such as mechanisms that facilitate human organ donation (Roth and Sönmez 2005, 2007).

Notes

1. The theory of mechanism design theoretically analyzes those mechanisms in which rational agents seeking their own interest with private information, generate the desired outcomes. In 2007, Leonid Hurwicz, Eric S. Maskin and Roger B. Myerson shared the Nobel Prize in Economics "for having set the basis for the theory of mechanism design."
2. Auction theory is born with William Vickrey (1961, 1962), who was also awarded with the Nobel Prize in 1996, although he never received it, because he passed away a few days after the award was announced.
3. Many of the main technology companies – Google, Microsoft, Hewlett-Packard, and Amazon – currently have economists in key positions. Although the scope of the ESD field is not precise, an important part of these economists' work is related to this area. An example of studies relative to this field is Varian (2007, 2009), and Reiley *et al.* (2010).
4. We could call ESD *microeconomic systems engineering*, in the sense that it is an applied branch which, starting from auction theory and mechanism design, together with the accumulated knowledge on empirical regularities, designs allocation mechanisms that subsequently use the laboratory as a test bed. For instance, if we compare it with aeronautic engineering, the latter uses theories and accumulated empirical knowledge from multiple disciplines – fluid mechanics, aerodynamic, propulsion, etc.- to propose concrete aircraft design. Roth (2002) proposes a similar analogy, with a slightly different focus.
5. All subjects value the good differently and each individual valuation is not related to that of anyone else. Each subject knows only his own valuation and how the valuations are distributed between all of them.
6. When $n = 2$ it is optimal to bid half of the private valuation ($b = 0.5 * v$), when $n = 3$ the optimal bid is $0.6 * v$, when $n = 4$ it will be $0.75 * v$, etc.
7. The sealed bid second price auctions have the name of William Vickrey (see footnote no. 2), who first studied them and theoretically highlighted their properties to incentivize the truthful revelation of private information. For a history of the practical applications of the Vickrey auctions, see Lucking-Reiley (2000).
8. The literature on auctions is very wide and extensive, and has explored different environments (one or multiple utilities for sale – with agents who demand one or multiple units –, with private values, common values, affiliated values, or common and private values, valuation asymmetries, information asymmetries, etc.), different mechanisms (FPA, SPA, English auctions, Dutch auctions, etc.) and diverse objectives (efficiency, auctioneer's income, truthful revelation, etc.).
9. The reader may consider that the exposition problem (both for complements and substitutes) can be eliminated by not allowing bids to be made simultaneously, but instead that goods are sold sequentially. Thus, the problem for the last good sold is eliminated, reducing it in an inverse direction to the order in which goods are sold, but it does not eliminate

the problem for the first goods being sold, which opens the question, which good should be sold first? – given that the order of the auction would affect outcomes. In addition, this may induce unwanted strategic behavior of participants. Klemperer (2002) shows some examples.

10. The first to investigate a combinatorial mechanism in the laboratory – and also to coin the term combinatorial auction – was Rassenti, Smith and Bulfin (1982, from now on RSB), when proposing a mechanism (which would allow one to bid by packages) to allocate rights over time slots for efficient landing/take off and limiting the exposition problem.

11. We can also see how two packages of slots may be substitutes: for instance, the JFK–Barajas package and the Newark–Barajas package could be substitutes for some airline – the degree of substitution will depend on the importance of the connections in each airport for that route.

12. Notice that with only three goods there are already seven options to evaluate per participant. The complexity of this type of problems increases exponentially with the number of goods: for n good there are $2n$ combinations. For the FFC auction in the U.S.A., where 2,074 licenses were being bid for the use of the spectrum, there were up to 22,047 possible combinations per participant (Porter and Smith 2006). This is known as the winner determination problem.

13. McCabe *et al.* (1991a, 1991b), in their attempts to test Vickrey's proposal of using English auctions with multiple units, found a jump-bidding behavior. Their results indicate that allowing participants to announce their bids in such a way that publicly shows their effect on prices is not a good design in settings with multiple units, because it allows impatient participants (or those who do not know how to refine their strategy) to make rough jumps in their bids and exceed the competitive price. In addition, it may facilitate collusion through the use of signals.

14. PRRS are neither the first nor the only ones who have designed and experimentally tested a combinatorial auction to reach an efficient allocation. Cramton *et al.* (2006) and Porter and Rassenti (2010) have also surveyed the literature.

15. Individual prices adjust to the demand of goods and/or packages, and they can even decrease for some individual goods belonging to packages where a certain good has an excess of demand. The mechanism can be more complicated: it is possible that together with the increase in price, a good may go from having an excess of demand to having an excess of supply – that is that "too many" people abandon the auction of a good – or that there are no competitive equilibrium prices for the individual goods and it is necessary to use pseudo-dual prices (see RSB 1982). In this case the non-dominated bids from previous rounds are included and a method of integer programming is used to find the optimal solution. In general, the auction that generates this allocation could be described as an algorithm to discover the upper bounds of the pseudo-dual prices. Porter *et al.* (2003) provide technical details of how the mechanism works in these cases.

16. The Simultaneous Multi-Round auction (SMR) is an auction with multiple rounds (with feedback between rounds) where bids for different goods can be simultaneously made, subject to certain activity constraints: continuous participation with minimal increments.

17. The CRA mechanism proposed a hybrid auction combining multiple rounds with continuous bidding periods, with activity constraints that imitated those in the SMR and others even more complicated. For more details, see Charles River and Associates Inc. and Market Design Inc. (1998a, b).

18. In addition to the theoretical predictions, there is experimental evidence on the strategic decrease of the revealed demand in auctions of multiple units with multi-unit demands (see, for instance, List and Lucking-Reiley, 2002). However, no evidence has been found on the decrease of the demand in combinatorial auctions.

19. In general, these have shown the potential that *forward* markets have to limit the generators' market power. For space considerations we will not be able to review these experiments in more detail.

20. They are typically faced with a regulated price that covers the average cost of generation, transport and distribution. However, it is common that the generation and transportation costs vary by an order of magnitude between the hours of maximum (peak demand) and minimum (base demand) consumption of the same day.

21. Or contracts, with dynamic prices, through which the consumption could be cut when prices are high.

22. Additionally, they use a radial network with three nodes connected through transmission lines with losses. This is not the most complex setting that has been used in experiments of electric energy markets. See for instance Olson *et al.* (1999).

4
Experimental Finance

Debrah Meloso and José Penalva

Introduction

If you have been reading the chapters of this book in order, at this point you are immersed in the workings of modeling and economic theory in general. We ask you to take one step back and picture a financial market. Think of traders sitting in front of computer terminals analyzing data and fancy charts; think of the million things they need to take into account, the complexity of it all; think of the thousands of transactions and the huge amounts of money changing hands every day, every second! And think that at the end of those trades there are people reacting to information and trying to deal with that complexity, with *risks*.

You have seen how people deal with risk before, back in Chapter 2. But now you will see risks as they relate to money, to the flows of capital that originate with firms' investment needs and consumers' borrowing requirements, and that are transformed through financial markets into complex structures that fluctuate and mutate as they change hands across the globe.

Let's give a concrete example: Jim has an excess of cash from an inheritance. He faces a great deal of risk as he evaluates the things he can buy with it, not just today, but over the course of his lifetime (let's not even talk about inflation). Jim considers different options:

- Investing in the high-tech start-up of a friend, Michael.
- Lending money to the local grocer around the corner, who has successfully run her business for over 25 years.
- Stashing the money under his mattress.

The first option links Jim's future consumption with the fortunes of Michael's ideas and management. Jim may also think that having all his cash in Michael's business is not a good idea. So he breaks up the inheritance amongst the different options, building a *portfolio* of investments.

In other words, Jim *diversifies* his portfolio of investments, assuming a little bit of risk (and return) from each of several different investments, rather than a lot of risk from one single investment. Investment portfolios, risk, and diversification are only a few of the basic ideas about finance that you are probably already familiar with. Let us look even further and assume that Jim believes that the three investment options considered so far are too few or unreliable. Jim can then turn to *financial markets*, where he will find lots of people already looking to trade, and where he can transact anonymously knowing that the institution (the market) ensures all deals are honored. The type of markets we usually think about and the type we will study in what follows are *centralized, transparent*, and *large*.

Experimental finance and financial markets

As we will see, experimental finance is a huge field of research to which we cannot do full justice in this chapter. We will focus on the study of (competitive) *financial markets*, for they provide the best starting point for delving into experimental finance. The study of financial markets, with its especially solid theoretical constructs, provides a great foundation both for understanding financial issues and the value-added of using experiments.

Markets facilitate capital movements by providing a place to exchange risks. Risks are exchanged in the form of financial *assets*: IBM stock has some risks associated with it, which are different from those of Apple stock. By combining one unit of each, yet another risk profile can be obtained, and so forth. We will study what experiments have to say about how people use markets and the asset prices that come out of them.

Then we will turn to the informational role of financial markets: we will look at what experiments say about the role of markets and prices in transmitting individual information to the whole of the economy, and what this does to people's incentives to acquire that information in the first place.

Finally we conclude with a very quick look at alternative motives for trading assets, the effect of behavioral biases, and a quick overview of other areas of experimental finance.

Risk-sharing and diversification

Returning to competitive financial markets: these are like regular competitive markets (as in Chapter 1) where people act as price takers and where the object of trade is not apples and oranges but promises of future payment (assets). These markets are implemented in laboratory experiments using *double oral auctions* (DOA) or open book markets (see section A in the Appendix), where trading is similar to what you see in regular electronic stock exchanges.

In the lab, these markets are opened for a fixed period of time during which experimental participants buy and sell freely. What participants trade are artificial assets and what we observe is the resulting prices, as in a regular stock market. Additionally, in the lab it is possible to observe whatever assets people hold before, during, and after trade (their initial, intermediate, and *final asset portfolios*). Final portfolios determine the payoff of experimental participants, when they are converted into cash by the experimenter.

It is important to understand why people trade and what the profits from trading are. As we will see, trading is based on differences between agents. We will focus on risk-based differences, but as you saw in Chapter 2, there are other factors that generate differences between agents: the way one weighs probabilities, how one values gains versus losses, how one computes probabilities, even whether in fact one uses probabilities at all. People also trade because they have different information, which raises additional issues which we will address later in this chapter.

There are two primary sources of risk-based differences:

a) Different *initial holdings*. Jim inherits IBM stock while Jenny inherits Apple stock. Jim and Jenny are identical twins, thus their tastes and perception of risk are the same, and their most preferred portfolios should look alike. To get to those portfolios they must trade Apple for IBM stocks.

b) Different *risk preferences*. Jim and Michael hold the same combination of IBM and Apple stock. However, the entrepreneurial Michael likes to hold more Apple than his more prudent friend Jim. They then benefit from trading and they end up with different investments in stocks (portfolios). Among the factors associated with differences in preferences are: age, health status, family situation, mood, etc.

These differences motivate trade in financial assets. When Jim exchanges exposure to one large risk for lots of small exposures to different risks, we say that he *diversifies*. In order to diversify using financial markets, Jim will have to find another person willing to exchange risky assets with him. In doing this, Jim and other agents involved in financial markets engage in *risk-sharing*. As you have seen in Chapter 2, agents' preference for risks is often to dislike them to a lower or higher degree – they are *risk averse*. Risk-averse agents will like to diversify and thus profit from risk-sharing in financial markets.

Before we discuss our first experiment, you need to meet a few additional important financial concepts. Suppose Jim wants to be able to travel the world with Michael if Michael's high-tech business goes really well. But he is also worried that his grandmother may need home care, as she gets older. To deal with this risk, Jim can invest part of his money in Michael's business, tying his fortune to Michael's, but also put part of his money with the grocer who will be able to pay him back if he needs the money for his grandmother. When Jim does this, he is taking into account that several eventualities or *states of the world* are possible in the future.

In our first example there are four states of the world: Michael's business may thrive, both if Jim's grandmother needs home care and if she doesn't (two states of the world); Michael's business may be a flop, whether Jim's grandmother needs home care or if she doesn't (another two states of the world).

If the assets traded in financial markets suffice for Jim to get any combination of payments across states of the world (in the example we need at least four assets), we say that *markets are complete*. To figure out how the states of the world affect the entire economy, we need to know the *market portfolio*. This is the portfolio of someone who (hypothetically) owns *all* the **risky** assets in the economy. In our example, the market portfolio includes all the shares in Michael's business as well as all shares in other risky stocks and investments.

Instead of investing in risky assets, Jim may want to invest in a *risk-free* asset, which is akin to a really safe bond. Cash acts like a risk-free asset, although usually risk-free assets offer a (very small) return.

Having seen what states of the world and the market portfolio are, you are finally ready to meet *aggregate risk*. This is the risk that cannot be eliminated by spreading one's investments across many assets (*diversification*) or across many people (*risk-sharing*), since it affects the market portfolio itself, which is the most diversified portfolio possible. Formally, there is aggregate risk if the payoff of the market portfolio is different in different states of the world. When we experience a "global" crisis it is because our real world has aggregate risk.

Risk-sharing experiments

We have seen that agents are exposed to great risks and prefer to share them. Markets provide a place where these risks can be traded, where it is possible to share them. There are a number of theoretical models that formalize the risk-sharing function of markets.

Experiments in this section are based on these models.

Consider a simple setting, where all participants have the same information, i.e. no participant has privileged (*private*) information. This allows us to focus on risk-sharing and its implications, and leads us to ask the following questions:

- Will participants trade to change their portfolios (diversify and share risks)?
- Do our models correctly predict the prices of assets and the final portfolios held by participants in the experiments?
- When markets are dynamic, will participants speculate (bet on price movements over time) and will speculation interfere with the profits of risk-sharing (reduce efficiency)?

A static financial market experiment

Bossaerts and Plott (2004) and Bossaerts, Plott and Zame (2007, from now on BPZ) report on a set of experiments aimed at testing the implications of the theoretical equilibrium models of risk-sharing. The main experiment, like all those in this chapter, has the structure presented in Table 4.1.

Table 4.1 Baseline structure of the financial market experiments in this chapter

---

1. Trading is done via an electronic stock market with an open book market (continuous double oral auction: continuous DOA, see Chapter 1 and Section A in the Appendix).
2. Participants are given cash and initial holdings of a small number of assets that they can trade in this market over a fixed period of time.
3. Participants are not assigned a role as buyer or seller. Instead, they can choose to buy, sell, or hold on to their initial holdings. *Short selling* (selling assets you do not already own) is not permitted.
4. Assets are entitlements to dividends (cash payments). Dividends are paid after trading and their exact value depends on the state of the world.
5. The probabilities associated with states of the world, as well as the relationship between the state of the world and asset dividends, are clearly specified and public information. We call this information the *distribution of dividends*.
6. The experimenter fixes and announces the distribution of dividends prior to the experiment. Participants can make profits from dividends but also from re-trading (buy cheap now and sell at a higher price later). Note that re-trade is a zero-sum game: what one participant wins another loses.
7. There is aggregate risk.

BPZ study one such financial market. In their experiment markets are complete and participants have different (and non-diversified) initial holdings, so we expect them to trade for the purpose of risk-sharing (see Section C in the Appendix for more details). Participants play over several periods in this experiment, and at the beginning of each period participants are "reborn" in an entirely new economy: prices, holdings, and trades that occurred in past periods do not affect initial holdings and the probability distribution of dividends in future periods (see Chapter 1).

The setup is *static*. This does not mean that participants can trade only once. In fact, the continuous DOA allows participants to trade and re-trade at their leisure (and we observe many trades). It is static because:

i. dividends are paid out only once; and,
ii. during a period participants acquire no new information about dividends.

At the end of each trading period, participants receive a payoff determined by their final holdings of assets and the dividends corresponding to the realized state of the world, which is revealed after the trading period ends.

The experiment is designed to capture important elements of the theory of *static symmetric information asset markets*. It is also designed to be able to compute indicators of the presence of risk motives for trade and to discern whether markets are in equilibrium.

We now cover the intuitive components of the theory and equilibrium notions. We consider three closely related models: the Arrow and Debreu model, the Radner model, and the Capital Asset Pricing model. We focus on the indicators and predictions that are tested in the experiments we review in this chapter. The interested reader will find the technical details of equilibrium and related indicators in Section B of the Appendix.

Equilibrium theoretical models considered here capture the following characteristics for markets where agents share risks:

- The world is risky. All risk in the world is contained in the risk of the market portfolio (aggregate risk). The risk of the market portfolio cannot be avoided through diversification.
- All risk-averse agents, trying to reduce their exposure to bad states of the world (states of the world where the market portfolio has low value), will bid up the value of money in such states: in bad states, money is scarce and sources of money in such states are thus expensive.

Although the economy as a whole does poorly in a crisis, some assets may actually do (relatively) well in a crisis (e.g., Wal-Mart or Carriage Services – a funeral services company). Since people are risk averse, if they think a crisis is very likely they will all want to hold assets such as Wal-Mart, thus bidding up their price. These assets are called *countercyclical* because their performance is negatively correlated with that of the market portfolio (when the market portfolio does poorly the countercyclical asset does well, and vice-versa).

To understand the results in BPZ you need a basic understanding of equilibrium concepts and related indicators. The first is the notion of equilibrium used to study the Arrow-Debreu economy (ADE), or the general competitive equilibrium (see Chapter 1).

Arrow-Debreu equilibrium: The ADE treats money in two different states of the world as two different products, like apples and oranges. In the ADE the price of an asset that pays one euro in one state of the world (and nothing otherwise) is equivalent to the price of money in that state of the world. To help visualize the states of the world, consider two states of the world, such as "Rainy" and "Not Rainy."

- If money is very scarce in the state of the world "Rainy" then it will command a high price – just as flawless diamonds are expensive because they are so rare.
- Nevertheless, if the state "Rainy" is very unlikely the price cannot be so high, since even risk-averse agents will pay little attention to a very unlikely (bad) outcome.

Thus, we measure the price of money in a state of the world using the state *price–probability ratio*: that is, the state price divided by the probability of that state. In Arrow-Debreu equilibrium, the price–probability ratio should be highest for states of the world where the market portfolio has the lowest value, and so on.

Radner equilibrium (RadE): RadE is used to study asset prices (instead of the price of money in different states of the world). RadE provides the link between the price of money in each state of the world (ADE) and the prices of assets (e.g., since money is expensive in poor states of the world, countercyclical assets will also be expensive). In this equilibrium, agents trade a given set of available assets, and agents' objectives are to achieve the *best* portfolio they can. The precise meaning of "best" is specific to each agent, but it always relates to the fact that agents are risk averse and need to optimize the trade-off between risk and gain.

A special case of the RadE model – the Capital Asset Pricing Model (CAPM) – assumes that the *best* portfolio is such that the trade-off between the expected value and the variance of the portfolio's payoff is maximized.[1] In equilibrium, the CAPM predicts that all relevant information about an asset's price and, consequently, its *returns* (an asset's *return* in a given state of the world is its dividend in that state divided by its trading price) is captured by the covariance between the asset's returns and the returns of the market portfolio.

More precisely, in the CAPM the equilibrium expected *excess returns* of an asset (the expected difference between the asset's return and that of the risk-free asset) are proportional to this covariance. Also, in the CAPM all agents will end up holding portfolios that look like small replicas of the market portfolio. This will imply that the market portfolio is *mean-variance efficient* in equilibrium, having the highest ratio between the expected excess return of the portfolio and its standard deviation (this ratio is called the *Sharpe ratio*).

In the experiment, BPZ look at several indicators to see if the predictions of the theory hold. In particular they look at:

- state price–probability ratios,
- the Sharpe ratio of the market portfolio, which is compared with the optimal Sharpe ratio (maximal Sharpe ratio that can be obtained given asset prices and payoffs), and
- the relative holdings of risky assets in the final portfolios of the experimental participants, to be compared with the market portfolio.

It is important to note that these indicators are easy to construct in the experiment because the experimenter knows the distribution of dividends and the market portfolio and he/she observes the portfolios of market participants. This is not so in the real world! BPZ's first findings are:

i. Price–probability ratios are ranked as expected: they are highest in the poorest state (X, see the Appendix), lowest in the richest state (Y), and the third state (Z) is between X and Y.
ii. The Sharpe ratio of the market portfolio converges to the optimal Sharpe ratio.

The first finding is consistent with the rank predicted in Arrow-Debreu equilibrium. The second finding (convergence of the Sharpe ratio of the market portfolio to the optimal Sharpe ratio) indicates that prices correspond to those expected in CAPM: such prices suggest that the market portfolio is mean-variance efficient.

Looking back over the questions that were asked about *risk-sharing experiments* at the beginning of the section we find that there is trade (trading volume is high) and prices display some of the properties predicted by the relevant models: state price–probability ratios are ranked as predicted (ADE) and the Sharpe ratio of the market portfolio is close to optimal (CAPM).

However, while prices in BPZ's experiments are consistent with CAPM, participants' final portfolios are not. The CAPM predicts that investors' final portfolios are a combination of the risky part of the market portfolio (scaled down, obviously) with some amount of the risk-free asset.[2] The third result of this experiment is thus:

iii. Participants in the experiment hold risky assets in proportions that are different from the proportions of the market portfolio. However, the mode of participants' holdings follows the market portfolio.

Because prices arise out of the trading which is necessary to attain final portfolios, and prices in the experiment are consistent with the theory, it is particularly surprising that portfolios are not. To explain this, BPZ develop a model where mean-variance preferences are only an approximation of investors' true preferences (see Bossaerts, Plott, and Zame, 2007, for a complete, but very technical description). In this new model, prices and the efficiency of the market portfolio are the same as in the regular CAPM, but equilibrium asset holdings are only on average proportional to the market portfolio.

Using this model, BPZ go back to the data and find that it is fully consistent with the new model.[3] This is a good example of the feedback that can arise between experimental research and the development of theory.

Replication

The BPZ experiment has been replicated with changes in the magnitude of aggregate risk, the type of asset correlations and the composition of the participant pool, and the same results hold. But you run into problems if there are too few participants. Nevertheless, with just three or four traded assets, price convergence is already fast and stable with as few as 20 participants.

Static vs. dynamic financial market experiments

The BPZ experiment is relatively recent and was run after a long series of experiments on financial markets with opposite conclusions, many of which were influenced by the seminal work of Smith, Suchanek and Williams (1988, from now on SSW; in Section D of the Appendix the reader can find a detailed description of the experiment). SSW find that prices diverge from their theoretical (*"fundamental value"*) levels – they find *price bubbles*.

The study of financial price bubbles is particularly relevant and interesting, so we will now look at some of the differences between the experiments (SSW vs. BPZ) to understand when bubbles may arise. There are three main differences:

1. The first and most evident difference is that SSW has a *dynamic* setup.
2. In the SSW experiments there is a single long-lived risky asset and cash, while in BPZ there are *several* risky assets.
3. Markets in SSW are not complete.

Consider this step-by-step. The SSW experiment has a *dynamic* setup: that is, participants and their asset holdings "live" for 15 *interconnected* periods. Periods are connected because the final holdings of one period (period *t*) are the initial holdings for the following period (period *t+1*). Also, in SSW an asset is a promise of payment at the end of *every* future period. That is, a participant that starts with and holds an asset during all 15 periods will receive 15 payments. A participant that buys the asset in period 3, buys the right to receive 13 dividend payments (two payments have already been made) plus the right of reselling the asset at any point in the future.

For a similar reason (four possible dividend payments in each out of 15 periods), markets are not complete: there are only two assets that can be traded while there are four *states of the world* in every period.[4]

In the SSW experiments there are two assets: cash and a *single* risky asset that lasts all periods. At the end of every period the risky asset has a 0.25% probability of paying one out of four possible dividends. The stream of dividends is independent and identically distributed (i.i.d.), meaning that the realization of a dividend in one period does not give any new information about the distribution of dividends at the end periods that follow. All this is public information.

In the BPZ experiments, agents hold and trade several *different* risky assets and by re-combining risky assets they can increase the efficiency of their portfolio, increasing the expected gain and reducing its variance. For example, Jim may have an initial portfolio containing only IBM stock while Jenny initially has only Apple stock. Even if Jim and Jenny were equally risk averse, they could both improve the efficiency of their portfolio by acquiring some of the stock held by the other. In that way, their final portfolios would imitate the market portfolio, which – as you have already deduced – is composed of both stock in IBM and Apple. On the contrary, in the SSW setting, with only one risky asset, participants' initial holdings are necessarily a fraction of the market portfolio (why?). The only thing left for agents to do is to decrease or increase the fraction of their total portfolio that is made up of the risky asset vis-à-vis the risk-free asset. If all participants are risk averse, this means that the more risk-averse participants will have to find the less risk-averse ones in order to achieve mutually advantageous trade. We therefore say that SSW provides *weak* reasons for portfolio rebalancing when compared to the BPZ multiple assets setup.

Given the differences we have mentioned, what would you expect to happen in the setting of SSW? Here is what happens.

Consider the *fundamental value* of the risky asset in SSW (see Section D in the Appendix). As time passes there are fewer periods (and payments) until the end of the experiment. In addition, there is no, and there will not be any, new information about dividends. Thus, the intrinsic value of the asset decreases and we expect prices to follow a similar path. Nonetheless, in the experiment we see the very robust appearance of price *increases* that are later drastically reversed in a crash (a bubble that later explodes). The appearance of bubbles is robust to many variations, including the imposition of a price ceiling or the restriction of participation to participants experienced with bubble markets in the past.

At the time SSW ran their experiment their results were taken to indicate that the positive experimental results found for competitive goods markets did not carry over to financial markets. We now know, as exemplified by BPZ, that this is not the main message of SSW. The main message is in fact that, in asset markets, convergence to equilibrium is very sensitive to the different layers of complexity of these markets. We use the three important differences between BPZ and SSW to explore this main message.

SSW propose, and provide evidence for, the hypothesis that speculation in a complex, dynamic environment (difference No. 1) is the
driving force for the appearance of bubbles. The main idea is that, in
order for prices to follow a path close to the asset's fundamental value,
participants need to understand how prices will behave in the future.
The experimental data show that, most probably, participants understand how the fundamental value evolves in time (in fact, prices do
crash back to the fundamental value in later periods). The problem
is that, even if a participant knows the evolution of the fundamental
value, she may think that others don't and, hence, that prices will
diverge from this value due to the irrationality of other participants.
Knowing this, a participant may participate in trade at the "wrong"
(irrational) prices, expecting to gain from re-trade to irrational participants at a later time.

Data on participants' price forecasts show that participants do not
base these forecasts only on the (well-understood) fundamental value
of the asset. Instead, they try to gage market irrationality and *adapt*
their forecasts to past forecast errors, supporting the above hypothesis. In general, this hypothesis is called a failure of *common knowledge of rationality*: even if you understand the fundamental value,
you believe that others don't and trade at wrong prices to exploit
this belief (see chapters 5 and 9 of Vol. 1). In other words, there is a
Pygmalion[5] effect in prices that makes them increase only to break
down in a big crash towards the end of the experiment, returning to
their "rational" level.

Lei, Noussair and Plott (2001, from now on LNP) provide evidence
that, even though the above hypothesis may influence the appearance of bubbles, it is not the main driving force. Instead, it is difference No. 3 that captures the essence of pricing bubbles. Let's see
how they show this. LNP replicate SSW in an environment where
participants have pre-assigned roles as buyers or sellers and, hence,
cannot speculate.

Without speculation there is no profiting from others' irrationality
over time and, hence, failure of common knowledge of rationality
cannot drive a pricing bubble. Nonetheless, LNP observe bubbles in
their setting! From this observation LNP elaborate their hypothesis
that bubbles are mainly due to *spurious* trade. This is trade that is
not motivated by risk sharing but by boredom or a feeling of duty in

the experimental setting.[6] Three observations strongly support this hypothesis:

- Bubbles appear in connection with large (unnecessary) trading volume.
- Trading volume falls and bubbles disappear when participants are experienced or when they are given a second task to undertake while trading.
- The LNP (and SSW) setup provides weak reasons for portfolio rebalancing.

In consequence, according to LNP, if the reasons for trade are not those assumed in the underlying theoretical model, there is no reason to expect prices to be as predicted by the model.

What about difference No. 2 (lack of complete markets in SSW)? Is it important? LNP also address this, by running a variation of the SSW setup where the risky asset pays one out of two equally likely dividends in every period.[7] Bubbles do not disappear, thus strengthening the relevance of difference No. 3.

We can now revisit the questions we posed originally in the *risk-sharing* experiments section, armed with the comparison across the three experiments (BPZ, SSW, LNP).

First, there is substantial trading volume, driven by risk-sharing motives (BPZ), by speculation, or – relevant only in the lab – by boredom (SSW and LNP). When driven by risk-sharing motives (as assumed in theory) this trade exploits gains from trade and leads to efficient outcomes. Thus – related to the second question – prices are consistent with theoretical predictions when the motives for trade are as assumed in the theoretical models. Theory's predictions on participants' holdings are supported to a lesser extent by the experimental data: the predicted correlation between participants' final portfolios and the market portfolio is observed for many participants but not all.

Regarding the third and last question, the results of SSW suggest that when the risk-sharing motive for trade is weak and, instead, the speculation motive takes over, speculation does indeed have a detrimental effect on efficiency. However, LNP show that speculation is certainly not the only thing responsible for the inefficiencies seen in a dynamic setting.

We end this section with a remark on experimental methodology. Notice that the theoretical relation between risk aversion and asset prices comes out strongly, provided that participants have a strong desire to hold final portfolios that efficiently trade off risk and expected return. This desire is hampered experimentally if:

- Participants are given efficient and well-diversified initial endowments, or
- Participants expect most of their payoff to come from initial cash holdings and not from their trading choices.

The above points relate to the *salience* property of experiments. An experiment is *salient* if the relation between experimental payoff and making the "right" choices is very strong. The experiments mentioned here show that salience is as relevant in asset-market experiments as in other experiments and that lack of salience hinders results. Whether an experiment has salient incentives or not can only be understood with the guidance of theory.

Informational asymmetries

A fundamental advantage of markets relative to other ways of organizing economic activity – such as central planning in the now extinct U.S.S.R. – is that they bring out information that would otherwise stay hidden.

Returning to Michael's high-tech start-up, Michael may understand very well the workings of his invention, but he may have very little information on relevant factors for its success such as whether there is a demand for his products. Meanwhile there are other agents, for example his potential consumers, who have much better information.

A centralized and transparent market allows everyone to credibly transmit their information[8] through openly observed prices, by "putting their money where their mouth is." We want to see what financial experiments have to say about this information revelation role of markets. We will now consider experiments where:

- The (prior) distribution of dividends is public information (as before), and so are prices.

- But we add the public information that *some investors may have private (privileged) information.*

In particular, some investors called *insiders* have more precise information about the distribution of dividends. One of the main theoretical notions in such economies is that of *Rational Expectations Equilibrium* (*REE* – see Section E in the Appendix). REE implicitly requires agents to do many complex calculations, as it assumes people use all their information in the best way possible, including the information embedded in prices. Also, REE generates very surprising and counterintuitive paradoxes, which we will now consider. Spoiler alert: a way out of the paradoxes is to use a modified version of REE – "noisy" REE.

Where is the paradox in REE and what does it have to do with insider trading? There are a few paradoxes. Let's start with the intuitive claim that an insider will be able to trade and profit from his privileged information, making *informational rents*. Suppose you are trading in an asset market and you are one of those with private information – an insider, like Gordon Gekko in the film *Wall Street*. If you try to buy, everybody else (paranoid that someone out there knows more than they do) will realize that you know the value is going to be high and the price increases. In fact, the moment you start buying, the price will shoot up, become too high for you to profitably trade, and reveal to uninformed traders that dividends are going to be high, that the price adjusts and there is little further reason for trade.[9]

We have just witnessed two paradoxes:

- Because of market paranoia the *insider* cannot profit from his privileged information (no informational rents), and
- prices reveal all information even though there is basically no trade!

These paradoxes are also important to good firm management. In theory, a bad manager will reduce firm value and the firm's share price will fall. A good manager can then buy the firm, fire the bad manager and increase the share price, which would greatly profit the good manager. But if shareholders realize that the firm is going to be taken over by a better manager, they will not want to sell at the low price. They will not sell until the price reflects the value the new manager is going to produce. But, then the price is too high and the new manager will not buy the firm.

Continuing with the paradoxes in REE, consider what would happen if private information in financial markets were not free but instead needed to be acquired with money or effort. In this case the above paradoxes lead to a third one:

- Without informational rents investors have no incentive to acquire information even if it is very cheap.

Thus, the informational efficiency of REE backfires: since nobody will acquire information for the market to aggregate efficiently, prices will reveal no information.

One may argue that none of the above effects will prevail if investors have more reasons to trade than just information. For instance, if you don't like your initial portfolio, you may trade in spite of the paranoia of markets with asymmetric information, because – at reasonable prices – you will still be interested in changing your exposure to risk (diversifying).

The problem is that REE has the power to eliminate even these motives for trade (and again, eliminate the informative function of the market). For example, if the private information of all insiders completely reveals the true state of the world:

- There will be no risk in a REE and, hence, risk-sharing motives for trade will not survive.

This is our fourth paradox, and a real *Catch 22*![10]

Despite these paradoxes, the movie *Wall Street* was a great success (in its time, 1987 – although you may know the 2010 sequel) not just because of the great acting, but because there are Gekkos out there, making money with private information every day, and they are doing just fine.

So where is the problem in the model and how do we change it to be more realistic? The code word used by economists here is "noise." We now turn to experiments where noise helps us find a way out of these paradoxes and improve the theory.

Almost all static experimental markets with private information follow one basic setup introduced by Plott and Sunder in 1982 (from now on PS). It is very special, because assets have *personalized dividends*: different participants receive different dividends in the same state of the world. This "trick" generates reasons to trade that are immune to all REE paradoxes. Even if the REE reveals the true state of the world, investors with a higher dividend in that state will be willing to buy it from those who have a lower dividend. This is not realistic, but it solves a problem that arises when studying "realistic" REE. In a completely informative REE there is no trade and, therefore, prices are not observed. How can we observe prices if nobody trades? Personalized dividends allow us to bypass the lack of reasons to trade, so that the experimenter can observe prices and so that holdings change in markets with asymmetric information. Moreover, this trick generates precise predictions as to who must hold the assets in REE: those investors who obtain the highest dividend in the state of the world that is revealed in REE.

In their experiments, PS assume markets have two or three states of the world whose probabilities are public information. Insiders are either told what the true state of the world is (*concentrated information*) or (in the three-state setup) they are told one of the two states that will *not* be realized (*dispersed information*). Meanwhile, other uninformed investors still believe that all three states of the world are possible.

Numerous variations on the basic PS setup (for instance, Forsythe and Lundholm, 1990) reveal that you can get convergence of market prices to REE levels, but that this convergence depends on several factors:

- First, concentrated information yields faster convergence to REE than dispersed information.
- Second, with dispersed information and (*ex ante*) incomplete markets, convergence occurs, but to an alternative equilibrium notion – one where investors use only their private information and ignore the information contained in prices.[11]
- Third, personalized dividends affect convergence to REE

Concretely, the third point states that convergence to REE is weaker as we go from (i) all investors have the same distribution of dividends, to (ii) investors know that the dividend distribution of others is one out of a few known options, and finally to (iii) investors know nothing about the dividend distribution of others.

Importantly, experiments reveal that, as predicted by theory, whenever there is convergence to REE, insiders do not make extra profits. We are then left with the question of what happens if information is not free – do agents acquire information?

Costly private information

Sunder (1992), and Angerer, Huber and Kirchler (2009), among others, add an information-buying stage before asset markets open in a PS setup. They ask whether information is acquired and, if so, whether REE ensues.

These experiments reveal that the key is whether the experimenter opts for a *fixed slots* setup or a *fixed price* setup. In the fixed slots setup, participants can competitively buy (via auction) a fixed number of "slots" for people to become *insiders*. In the fixed slots setup, perhaps unsurprisingly, as periods progress, investors notice that gains from private information are very low, so that insider slots are auctioned at ever-lower prices. REE emerges because, regardless of how little insiders pay for information, there is always a fixed number of insiders, so that markets remain informative and efficient.

In the fixed price setup, any number of investors can buy information and become an *insider*, as long as they are willing to pay a fixed (low) price. In the fixed price setup, with gains from information close to zero, the number of information buyers, given a fixed price, should go to zero, and, therefore, the market should become uninformative and inefficient. But in the experiment it doesn't because gains from information are always kept sufficiently high to compensate investors for the price paid for information.

How is this possible? Recall our "spoiler alert": noisy behavior (and noisy REE). In the fixed price setup there is unpredictable variation in the number of insiders and in their ability to act on this information. Experimental results portray this variation. This is *noise* that makes other investors uncertain about whether prices are informative or not. This doubt allows insiders to extract some rents (killing one of the paradoxes), which keeps the information flow in markets.

Thus, we observe prices that reflect private information with noise (where the noise is in the minds of non-insiders, who are uncertain as to how much information may be implicit in prices), and this is captured theoretically by the notion of *noisy rational expectations equilibrium* (NREE).

The above analysis of information aggregation in markets was done with experiments in which all private information was concentrated and, therefore, an investor could either be informed (if he bought *all* information) or uninformed. Other experiments have looked at the paradoxes in a context where investors can acquire information of different quality. The first results in this line of research show that only the insiders with the highest quality information can reap informational rents, while other insiders are worse off than investors who choose to buy no private information whatsoever.

Given the importance of noise to the study of information acquisition and trading, some experimenters have introduced it explicitly. Doing this helps them study how noise affects the trading behavior of investors with different market platforms (*protocols*) – other than the standard electronic stock market (DOA). The corresponding notions of equilibrium are very specific and beyond the scope of this chapter.

Other reasons for trade

As we mentioned in the introduction, there are many reasons to trade in financial markets: differences in aversion to losses, ambiguity, differences in mood, and in computational ability. It is *per se* interesting to wonder how these differences can lead to trade that is meaningful (our first question for all experimental financial markets). Even more interesting is to wonder whether, as in REE, the reasons for trade are "eaten" away in equilibrium. Do behavioral biases and computational limitations wash out? Do we observe efficient prices and (portfolio) holdings in equilibrium?

Camerer (1987) addressed this question for biases related to Bayesian updating (see chapter 2 of Vol 1) of public information in dynamic markets. Surprisingly, he found that individual difficulties with Bayesian updating rarely appear in prices (see Section F in the Appendix for the precise updating problem) – just as information differences disappear in REE.

As for ambiguity aversion we find that, on the one hand, individual biases related to risk measurement (e.g., Bayesian updating biases) transform into ambiguity aversion in a market context. On the other hand, ambiguity aversion may not wash away in equilibrium, neither theoretically nor experimentally (Asparouhova et al. (2015); Bossaerts et al. (2010); Section F in the Appendix). A theoretical overview of the issues that may drive the market effect of biases in (financial and non-financial) markets is given in Fehr and Tyran (2005). In particular they find that the key is whether the presence of biases drives non-biased agents to either simulate them (so that biases will be reflected, even exaggerated in prices), or exploit them (eliminating the bias in prices).

Real-world markets are of course complex brews, in which all imaginable motives for trade converge. The power of experiments is not that they realistically replicate these markets, but rather that they isolate the different forces involved in financial markets. In this way, each force's relevance and the quality of models based on them can be understood. What comes out is a better set of lenses with which to look at real-world markets.

Conclusions

We conclude by referring to areas of *Finance* that we have left out of this chapter (see Figure 4.1). We have focused our attention on a couple of aspects related to financial markets and the prices that emerge (*asset pricing*). But finance is more than a bunch of highly paid traders staring at computer screens. Finance encompasses *the study of all existing and potential mechanisms that economic agents use to raise and allocate capital*. Individuals and firms must decide how much they will spend and where they will obtain the funds to finance their expenses and investment projects; similarly, they must decide how to save to "make their money grow" or protect themselves against an uncertain future.

Financial markets describe the interaction between those needing finance and those wanting to invest. But investment and borrowing is not just about share prices and diversification.

Other sub-fields of finance study how financial markets or particular asset structures may emerge, how those markets are organized, and the purpose of the variety of financial actors we observe in financial markets (financial engineering and market microstructure). Most actors are intermediaries of some sort, agents that intermediate between those who can use money for different enterprises and those who have that money (see Section G in the Appendix). Among the many such actors we find banks, insurance companies, investment banks, funds (investment, hedge, and mutual funds), market makers,

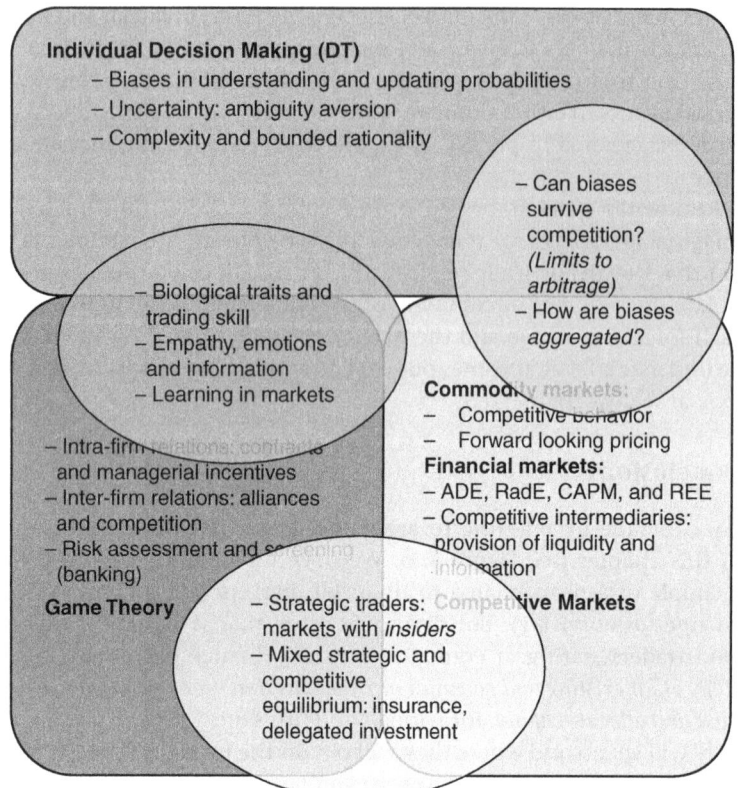

Figure 4.1 summarizes the topics of interest, showing for each case how they overlap with topics covered in other chapters in this book (sometimes they overlap with more than one topic at a time). As you can see, the field of Finance is large! And there exists experimental research relevant to most subfields of finance.

brokers, and (now) trading algorithms. A great deal of effort and research is dedicated to how those markets should function and be regulated (especially in banking and insurance).

Last, but most definitely not least, the decisions firms make about how to finance their expenses and invest their excess cash flows have implications for firm structure and are in turn affected by firm organization. For example, a firm that decides to have its shares publicly traded will often be "owned" by persons who may have very little idea which firm's stock is in their portfolio, let alone how a firm should be managed! Hence, the study of a firm's finances is also concerned with the relationship between stockholders and managers, the internal organization of the firm, and even with the firm's relations with its competitors and allies. These are all participants studied in *Corporate Finance*.

Appendix

A. The open book or double oral auction (DOA)

As discussed in Chapter 1, this is a trading protocol where all participants are allowed to buy or sell assets. Participants trade directly with each other by submitting orders, which are matched to each other through an "open book." An open book is a book of orders listing all the trading orders available for execution/trade, organized by time of arrival and execution price, and visible to all participants. There are two basic types of orders: limit orders and market orders.

Limit order

A limit order to buy specifies an amount and a maximum price a trader is willing to pay to buy that amount of the asset, e.g. "buy 1,000 shares of Google at £625 per share." A limit order to sell specifies an amount and a minimum price a trader is willing to accept to sell that amount of the asset. Limit orders that are not immediately executed build the order book, where they are listed while they wait to be executed.

Market order

The counterpart to a limit order. It asks to buy (or sell) a number of shares at the best possible price, e.g. "sell 1,500 shares of Google." It executes by matching up with existing limit orders. Since this order immediately transforms into a trade, it does not build the order book.

Bid. An order (limit or market) to buy.

Ask. An order to sell.

Trading protocol

A description of the set of rules that regulate trade. The trading protocol specifies the information displayed to each participant (such as whether they can see only the best price to sell and to buy, or the whole order book, or just the best three orders in the book, etc.) the type of offers participants can make, and the way in which participant behavior is allowed to evolve over time. Here we consider only the double oral auction or open book market, but many protocols are used in the real world and in experiments. An important open question is: Why do we observe so many different protocols?

Technology and experimental DOAs

Vernon Smith ran his first experimental DOA (Smith 1962) by building the book on a chalkboard, and keeping track of executed trades in a hand-written notebook. Smith, Suchanek and Williams (1988) used a version of software introduced by Williams (1980) for electronic trading (PLATO). PLATO was well equipped for trading one asset, less so for multiple or correlated assets. Since then, many new electronic DOAs have been developed, including Plott's MUDA, jMarkets (Advani et al. 2003), and the more recent Flex-e-Markets (http://www.flexemarkets.com). DOAs can also be implemented with the flexible experimental programming software z-Tree (Fischbacher 2007).

B.   The static setup – theoretical background for experimental data analysis

There are two dates: today and tomorrow. Today investors trade to get from their initial asset holdings to the best possible final asset holdings – they reorganize their asset portfolios. Tomorrow they find out what the state of the world is and consume the money they obtain from the assets holdings in their portfolio.

Arrow-Debreu equilibrium

In an Arrow-Debreu economy, investors can trade Arrow-Debreu (AD) securities. There is a single commodity ("money"), and one AD security for each state of the world – it pays one euro in that state and zero in all other states. So, for example, suppose there are two states of the world: Rainy and Sunny. The AD economy has two securities: holding one unit of the Rainy security provides one euro if it rains and nothing otherwise; holding one unit of security Sunny provides one euro if it is sunny and nothing otherwise.

An Arrow-Debreu equilibrium, ADE, specifies prices of AD securities $(p_1, \ldots, p_s)$, and holdings of these securities for each investor, such that: i) given prices, each investor maximizes the expected utility of money in each state of the world, and ii) demand for money equals supply of money in each state of the world.

Interpretation

The study of AD economies isolates the effect that preferences (especially risk aversion) will have on all asset prices. The price of an AD security is the value (today) of guaranteeing one euro in a given (future) state, relative to guaranteeing one euro in another (future) state of the world.

Ranking of state price probability ratios

If agents are all risk averse, it is more valuable to guarantee one euro in a poor state of the world (where the supply of money is small) than in a rich one (where there is lots of money). This means that prices of AD securities divided by the probability of each state of the world, satisfy:

$$\frac{p_s}{\pi_s} > \frac{p_{s'}}{\pi_{s'}} \Leftrightarrow W_s < W_{s'}$$

where s and s' are two states of the world, W is total amount of "money" available, p stands for price, and π is probability.

Radner equilibrium

In a static *Radner* economy investors can trade any number and type of assets. As in the Arrow-Debreu equilibrium, investors also maximize the expected utility of money in future states of the world, but do so indirectly, by constructing portfolios of assets whose payoffs will determine money received in future states. The difference is that assets are more complex and generally they are not AD securities. In this economy, an asset is more valuable if it provides investors with a useful distribution of payments in future states.

A Radner equilibrium (RadE), specifies prices of traded assets $(q_1, \ldots, q_K)$, holdings of assets and final consumption (of money) in each state of the world for each investor, such that: i) given prices and asset payoffs, each investor maximizes the expected utility of money in each state of the world, which he obtains from his assets, and ii) demand of each asset equals its supply (sum of initial holdings).

Pricing kernel

If the economy has AD securities, their prices can be used to price all traded assets. That's why these prices are also called the "pricing kernel." In a general RadE the price of an asset is the expected (discounted) value of its payoffs. The expected value is calculated using probabilities that are proportional to the price of (hypothetical) AD securities. Therefore, if we assume a discount factor equal to one, in a RadE with implied AD prices ($p_1, \ldots, p_s$ normalized so they add up to one), an asset with payoff vector $D = (D_1, \ldots, D_s)$ must have the following price q:

$$q = p_1 \times D_1 + p_2 \times D_2 + \cdots + p_s \times D_s$$

In complete markets you can obtain the implied AD prices in this equation from observed asset prices. This implied pricing kernel must satisfy the same ranking property as prices in an ADE (ranking of state price probability ratios). This is a result that is used for data analysis in experimental financial markets.

CAPM

The Capital Asset Pricing Model, CAPM, is a special RadE where investors only care about the *mean* and the *variance* of the distribution of money in future states of the world: that is, investors care about the *mean-variance efficiency* of their portfolios. For each state of the world, define an asset's *return* as its future payoff (dividend) divided by its (current) price. A portfolio is mean-variance efficient if no other portfolio can deliver the same mean future return with a lower variance of returns.

An asset whose return is always the same, regardless of the state of the world, is called a risk-free asset and its return, the risk-free return. The Sharpe ratio of a portfolio is the ratio of the difference of its mean return from the risk-free return, divided by the standard deviation of its returns. It is used to measure mean-variance efficiency. In every CAPM economy, we can compute the maximal achievable Sharpe ratio, which is:

$$\sqrt{\left(R - R_F\right)^T \Sigma^{-1} \left(R - R_F\right)}$$

where R is the vector of mean returns of all assets, and Σ is the matrix of asset return covariances. R_F is the risk-free return rate.

In a CAPM equilibrium the market portfolio is mean-variance efficient. Hence, its Sharpe ratio equals the above maximal Sharpe ratio. This is an important measure of convergence to equilibrium in the experiments we study.

C. Experiments

Bossaerts, Plott and Zame, 2007

The experiment encompasses nine sessions. The examples below are computed with parameters of period eight of session 011126: 36 participants, 18 of type I and 18 of type II. There are three states of the world (X, Y, and Z, with probabilities 0.46, 0.27, and 0.27 respectively). Two risky assets, one risk-free asset (*Notes*) and cash.

Table 4A.1 Dividend distribution of assets in francs (F, experimental currency)

	State		
	X	Y	Z
Asset A	170	370	150
Asset B	160	190	250
Letters	100	100	100

Table 4A.2 The initial asset endowments

	A	B	Letters	Cash(F)
Type I	5	4	−22	400
Type II	2	8	−23.1	400

From Table 4A.2 we learn that the per-capita *market portfolio* is given by 3.5 units of security A, and six units of B. No single participant holds the market portfolio; hence, participants are differently exposed to risk – there is *idiosyncratic risk*.

We combine Tables 4A.1 and 4A.2 to construct the distribution of money across states of the world implied by the market portfolio (Table 4A.3). Clearly there is a lot of aggregate risk. State X, with

Table 4A.3   Earnings of the market portfolio in each state of nature

	The market portfolio
X	$170 \times 3{,}5 + 160 \times 6 = \mathbf{1.555}$
Y	$370 \times \ 3{,}5 + 190 \times 6 = \mathbf{2.435}$
Z	$150 \times \ 3{,}5 + 250 \times 6 = \mathbf{2.025}$

the lowest payoff, is a *crisis*, while Y is a state of bonanza and Z is stuck in the middle. (How should state price probability ratios be ranked?)

We give an example of how the state price–probability ratios of Figure 4A.1 are computed using period 8 of session 011126. We use the average trading prices of the last ten trades to compute state

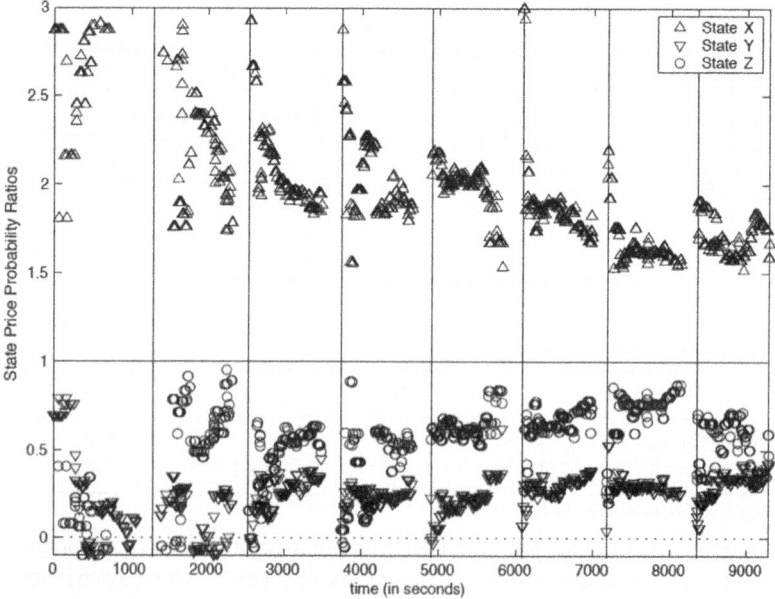

Figure 4A.1   State price–probability ratios

The price–probability ratios are ordered in the direction of the theoretically predicted ranking

Source: Bossaerts and Plott (2004)

price probability ratios: $q_A = 190F$ (price of security A), $q_B = 175F$, and $q_N = 100F$. Remember (Section B) that the price of security A in a RadE satisfies:

$$q_A = p_X \times D_{AX} + p_Y \times D_{AY} + p_Z \times D_{AZ}$$

where p_x is the state price of state X and so on, and D_{AX} is the payoff of security A in state X and so on. This yields a system of equations (in Table 4A.4) that we can solve to find the values of $p_x = 0.76$, $p_y = 0.11$ and $p_Z = 0.13$. We find the state price–probability ratios, p_x, by further dividing state prices by state probabilities, as computed in Table 4A.4. The state price–probability ratios are:

Table 4A.4   State price-probability ratios

• Px = 0,76/0,46 = 1,65,	190 = 170Px+370Py+150Pz
• Py = 0,11/0,27 = 0,41,	175 = 160Px+190Py+250Pz
• Pz = 0,13/0,27 = 0,48.	100 = 100(Px+Py+Pz).

Again, using data from period 8 of session 011126, we give an example of how to compute the Sharpe ratio differences represented in Figure 4A.2. First, compute the securities' returns using final prices: divide the first row of Table 4A.1 by $q_A = 190$ to obtain A's returns in each state of the world. Do the same for the other two securities. Then, use probabilities to obtain mean returns: $R_A = 1.15$, $R_B = 1.1$, $R_L = 1$, the variances and the covariance: $\sigma_A^2 = 0.24, \sigma_B^2 = 0.045$, and $\sigma_{AB} = -0.013$. Then, $\Sigma = \begin{bmatrix} 0.24 & -0.013 \\ -0.013 & 0.045 \end{bmatrix}$, and the optimal Sharpe ratio is 0.598 (using the formula in Section B). To compute the *Sharpe ratio* of the market portfolio we first compute its price: $p_m = 3.5 \times 190 + 6 \times 175 = 1715$. We use p_m and the dividends in Table 4A.3 to obtain the market portfolio returns: 0.91 (X), 1.42 (Y), and 1.18 (Z). Then compute the mean, the variance, and finally the Sharpe ratio of market returns, which is 0.555. It is very close to the optimal Sharpe ratio!

D.   Smith, Suchanek and Williams 1988

The experiment consisted of 27 sessions, all of them slight variations of each other. We report on session (28x; 9): nine experienced

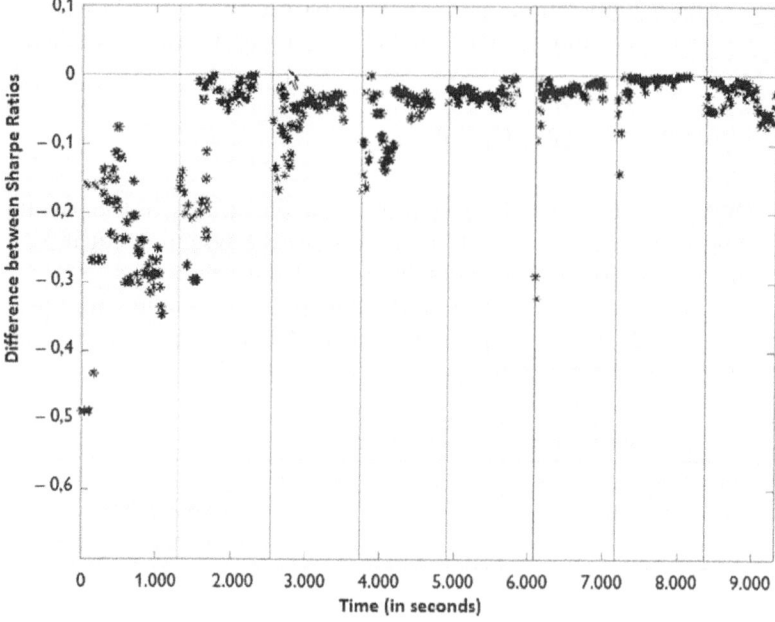

Figure 4A.2   Difference between the *Sharpe ratio* of the market portfolio and the optimal *Sharpe ratio*

Source: Bossaerts and Plott (2004)

participants, three of each type – I, II, or III. Fifteen trading periods with dividend payments at the end of every period, four states of the world per period (states s1, s2, s3, and s4, with probability 0.25 each, independent draws across periods). One risky asset plus cash.

The distribution of dividends given in Table 4A.5 is identical in every period, independent of draws in past periods. Notice that the

Table 4A.5   Dividends of risky assets in each state of nature in a single period (in cents)

	State			
	s1	s2	s3	s4
Dividends				
	0	8	28	60

Table 4A.6 Initial allocations by type of subject

	Type of subject		
	I	II	III
Risky	3	2	1
Cash	$2,25	$5,85	$9,45

distribution of *total dividends* changes in time. At time *t=0* there are 4^{15} states of the world, and this number falls at a ratio of four per period as dividend realizations are revealed.

The riskiness of each type's initial endowment is different (Table 4A.6). Notice that – unlike in the multiple-assets case (BPZ), with one single asset – a participant cannot improve his portfolio in both mean and variance. If he increases the mean expected return, the variance will increase, while if he reduces variance he will also reduce the expected return. There is always a trade-off, which is settled depending on the participant's risk aversion.

What about risk aversion? As we saw for the BPZ experiment, an asset's price is not usually equal to its expected payoff. We should therefore question whether the expected sum of dividends is a good reference point to use as the fundamental value of the asset in SSW. The answer is yes! The reason is that since there is only one risky asset, the price of the asset in a market populated with risk-averse participants can never exceed this expected sum of dividend pay-outs. If SSW observed bubble-shaped behavior of trading prices that, nonetheless, remained always below the expected sum of dividend pay-outs, then their results would be questionable. A bubble-shaped evolution of prices where prices move much above the expected sum of dividend pay-outs is incompatible with equilibrium theory and, hence, a real price bubble!

In Figure 4A.3, trading prices are shown as dots connected with solid lines, bids are shown as solid dots and asks are empty dots. The bubble is visible as a large departure upwards from the asset's fundamental value (expected dividend pay-out remaining), shown as straight solid lines. This fundamental value is necessarily decreasing in time, as the asset has ever fewer remaining dividend pay-outs and the expected value of these dividends does not change with time.

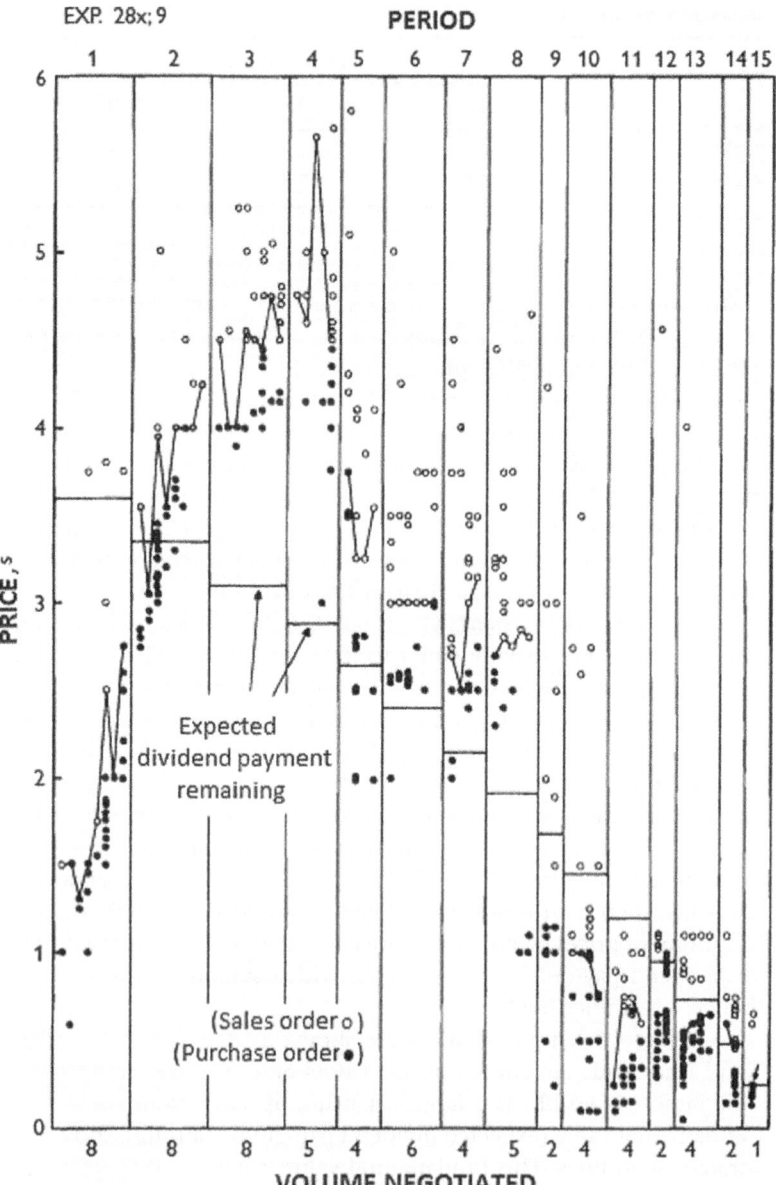

Figure 4A.3   The bubble of session (28x; 9)
Source: Smith, Suchanek, Williams (1988)

We can compute this fundamental value: the expected dividend in a single period is 24 cents. This means that at $t = 0$ the expected dividend pay-out remaining (15 periods: $15 \times 24 = 360$ cents) is \$3.60. We can compute this fundamental value for every other period by subtracting 24 cents per period. We needn't know the draw of dividend in past periods to compute this! For example, after five periods with their corresponding dividends, there are ten periods remaining and the asset's fundamental value is \$2.40, regardless of what the dividends were in the first five periods.

E. Rational expectation

An investor in a market has *rational expectations* if: when he observes public information he does not regret choices he made in the past – these choices may themselves be part of the public information. In financial markets where there is no private information, *rational expectations* matter in *dynamic markets*. With private information, *rational expectations* matter in both the static and the dynamic case.

In a *rational expectations equilibrium* (REE), we assume all investors can back out private information implicit in prices when they observe those prices. Armed with this power, the REE specifies prices and asset holdings: such that, i) investors maximize expected utility given their beliefs, ii) investor beliefs incorporate all information they can back out from prices, and iii) demand equals supply of assets.

Public information – dynamic

In a setup like that of SSW, demand in one period depends on the distribution of dividends and on the beliefs that investors hold about future prices. If these beliefs are wrong, they will regret the demand they submitted in the past. However, their own demands helped determine the price that ultimately proved them wrong. Only with rational expectations will investors submit demands that generate prices equaling their beliefs. No investor then regrets their choices.

Private information

Example: three states of the world, there is an *insider* who knows only two of these states may occur, the other one will never occur. The outsider submits a demand function based on her belief that all three states are possible. Since she knows there is an insider, once she sees the price, she deduces what information this insider had and regrets her demand.

In a *noisy rational expectations equilibrium* (NREE) the relation between price and private information is probabilistic, so investors cannot perfectly deduce information from prices. Everything we said about REE remains the same here, except that even after seeing the price, investors do not perfectly know what information was in the markets. *Noise* can be motivated in many ways: there are traders that trade "irrationally," blurring prices for everyone else; or, markets are participant to unpredictable supply shocks; or, states of the world do not capture all existing uncertainty.

F.   Behavioral bias in markets

Bayesian updating in laboratory financial markets

A large proportion of participants have problems applying the law of probability named Bayesian updating. Consider the following experiment (Camerer 1987): There are two states of the world, X (prob. 0.6) and Y (0.4). There is an urn with three balls in it: In state X this urn contains one red ball and two black balls; in state Y it contains two red balls and one black. At the beginning of the experiment, the experimenter takes out three balls (the *sample*) from an urn (with replacement: takes one out, shows it, and puts it back in the urn) and shows them to everyone. The experiment consists of verifying how people use the information from the balls drawn at the beginning (the sample) when they can trade financial assets. Participants trade a risky asset with dividends of 600F in state X and 200F in Y.

If the three-ball sample is "two red and one black," what is the probability of state X?

$P(X|2 \text{ reds}) = P(X \& 2 \text{ reds})/P(2 \text{ reds})$. We use Baye's rule to find that $P(X \& 2 \text{ reds})=0.133$ and $P(2 \text{ reds})=0.311$, implying that $P(X|2 \text{ reds}) = 0.428$. We can now compute the expected value of the risky asset after seeing a sample with two red balls: *E(asset value|2 reds)* = 0.428x600+0.572x200=328.6.

Theoretical bias – exact representativeness

Notice that a sample of two red balls and one black ball exactly matches the contents of the urn in state Y. A participant displays exact representativeness if she wrongly believes that *P(Y|2 reds)* is close to *one*, when in reality it is 0.572. That is, after observing a sample that matches one of the possible populations exactly, the participant assigns probability close to one to this population. The paper finds support for widespread use of exact representativeness, but it does not show up significantly in asset prices.

Asparouhova et al. (2015) analyze the Monty Hall problem, another bias in Bayesian updating based on a TV show (Google it, it is very cute!). They find that in a market context this bias delivers prices and holdings undistinguishable from ambiguity aversion.

Ambiguity aversion in laboratory financial markets

Ellsberg's one-urn paradox in markets (Ellsberg 1961, Bossaerts et al. 2010): Consider a similar setup as the Bayesian updating ones above. An urn contains red, green, and blue balls. There are nine balls. Of these, three are red and the rest are either green or blue, in unknown proportions. Balls are drawn at the end. Participants can trade three AD securities: security *Red* pays \$0.5 if a red ball is drawn and zero otherwise. Similarly for securities *Green* and *Blue*. We call green and blue the *ambiguous* states, since their exact probabilities are unknown.

You are asked whether you prefer security *Red* or *Green*. If you are a median person, you say *Red*. This means you think that *P(state red)* = 1/3>P(state green). Next, you are asked whether you prefer a portfolio of one *Red* and one *Blue* or a portfolio of one *Green* and one *Blue*. Being a median person, you choose the second portfolio, expressing that you believe *P(state red) + P(state blue) < P(state green) + P(state blue)*, which is inconsistent with your previous choice. This is because the median person is *ambiguity averse*: you dislike betting on objects for which you don't know the exact probabilities. Ambiguity-averse participants, in this experiment, are less reactive than other participants to changes in prices. Their "stubborn" desire to hold a certain type of portfolio, regardless of asset prices, may cause an endogenous change in the market portfolio that is left over for trade among the ambiguity neutral participants. The new (endogenous) market portfolio implies a new ranking of state price–probability ratios. When the authors (Bossaerts et al. 2010) find that state price–probability ratios are ranked according to the endogenous market portfolio, instead of the original one, they find evidence that ambiguity aversion is present and persistent in financial markets.

G.   Financial actors

Investors: These financial actors start out holding either cash or a set of assets that they wish to trade into a final portfolio of assets. Based on the distribution of dividends they trade to *control the future payments they will receive*.

Firms: These are complex structures that appear in finance at many levels. They raise capital in financial markets, thus creating *risky assets* for investors to trade; they decide how to finance their operations (debt, capital raising); and they resolve complex contractual problems between owners and employees and across firms.

Managers and stockholders: In the publicly traded firms that are the focus of most of Finance, stockholders own the firm and delegate its operations to a manager. Managers decide what projects to pursue. Their incentives are not always aligned with those of stockholders. The resolution of this conflict is attempted through contracts.

Banks, investment funds, and investment banks: These are all intermediaries that screen and package risky investments. They either bear some of the risk and screen investment opportunities (banks), or they create products that are sold directly (funds) or in competitive markets (investment banks).

Market makers: Intermediaries that trade in financial markets. They accumulate inventories in high supply periods and deplete them in high demand periods. In this way they make markets more liquid. Although they trade in markets they do not have the same (consumption) motivation as investors. They are motivated by the premium that traders are willing to pay in order to trade in a timely manner.

Notes

1. In general, an agent could care about expected value, variance, skewness, kurtosis, and even the shape of the entire distribution of returns. CAPM assumes they only care about expected value (positively) and variance (negatively).
2. An investor with mean-variance preferences will optimally hold a portfolio with maximal Sharpe ratio, and in equilibrium, the market portfolio has maximal Sharpe ratio.
3. Can you think of an explanation as to why experimental asset holdings differ from the theoretical CAPM ones but not prices? To do this exercise you need to understand the contents of Section B in the Appendix very well.

4. We remain superficial about the notion of market completeness since a more detailed discussion would require a long detour from our main topic. It suffices to know that for a market to be complete there must be at least as many traded assets as there are states of the world.

5. The Pygmalion effect comes from a poem by Ovid (*Metamorphosis* X) where Pygmalion the sculptor sculpts a female figure with such beauty (Galatea) that he falls in love with her and treats her like a real woman. Aphrodite has mercy on him and turns the statue into woman.

6. Participants in the laboratory feel "obliged" to do something.

7. We have not talked so far about the meaning of *complete markets* in a dynamic setting. It is more complicated than for static markets. Still, the LNP treatment with only two possible dividends is almost surely a complete markets setup.

8. In Chapter 9, third section, a similar idea is discussed: voting as a mechanism of information aggregation.

9. Symmetrically, if prices don't shoot up in spite of your efforts to buy, you may get suspicious that your private information was inaccurate! This is also an important part of the story that ultimately may paralyze markets with asymmetric information.

10. Catch 22 is a very good anti-war novel by Joseph Heller. Jose and Debrah highly recommend it!

11. *Ex-ante*, that is *before* any private information is transmitted. Private information can reduce the number of relevant states of the world and, thus, transform a previously incomplete market into a complete one.

5
Financial Crisis and Panic in the Laboratory

Hubert János Kiss, Ismael Rodriguez-Lara and Alfonso Rosa-Garcia

Introduction

Thousands of depositors crowd in the door of a bank branch. "Where is our money?" they shout enraged. "We want our money back!" The scene, besides being part of the classic movie "It's a wonderful life" (Frank Capra, 1946), reflects a reality that many people believed distant but that has re-arisen strongly in recent years. We refer to bank runs.

When one thinks about economic crises, it is not hard to associate them with bank runs. Bernanke (1983) in fact argues that these were the cause of the major economic losses that took place in the Great Depression during the decade of 1930. More recently, many experts have spoken about the massive withdrawals of deposits in September 2007 from the British bank Northern Rock as the event that announced what has become known as the Great Recession (2008–2015). Whether or not that event marked the beginning of the current crisis, the massive withdrawals from this bank illustrate that bank runs, as well as financial crises, are neither events of the past nor isolated phenomena. They can take place today in advanced economies. In this same way, many countries have suffered similar problems in recent years. Whether it is the case of the Bank of East Asia, of Bankia in Spain, or that of American entities such as Mutual Washington, Bear Stearns or Lehman Brothers, in all them sudden money withdrawals of massive amounts have occurred during the recent financial crisis.

Our aim in this chapter is to talk about financial crises. Our intention is to complement the discussion in Chapter 4 on Experimental

Finance, where the reader can familiarize themselves with the problem of bubbles and market risks, but with a different approach. In this chapter we will study the behavior of the depositors and investors during a financial crisis and will analyze (theoretically and experimentally) what can lead a depositor to withdraw his money in moments of crisis. In addition, we will discuss the literature on informational cascades and will see the mechanisms that may lead to the contagion of panic. In this sense, knowing how information is transmitted plays a key role, since there is no doubt we live in a globalized world, where both depositors and investors can have information on almost anything that is happening. Our chapter will analyze, therefore, if there is an effect, and how it takes place, by observing what others have done from an individual point of view (will I withdraw my money if I see that others are doing so?), as well as from a more global point of view (will it affect depositors and investors to know what is happening in a bank nearby or in a similar fund?).

Before beginning, we would like to point to three things. The first is that we will center our discussion on bank runs, even though the reasoning and analysis can be extended to different financial areas. In fact, many of the studies that we are going to analyze argue that their findings are generalizable to other markets and institutions that also fail during financial crises. For example, investment and pension funds, the repo market, markets for interbank loans or the stock market. That is why massive withdrawals of money, information transmission and contagion problems can all take place in these contexts, affecting even the economic stability of nations.[1] Secondly, though the topic that we treat could seem to be eminently practical, it is our intention to briefly introduce some theoretical notions that clarify the problems at hand. To do so, we will discuss some theoretical predictions of bank run models and herd behavior models, before presenting the experimental evidence. The presence of a significant amount of experiments is, in fact, the third point that we would like to introduce. Even though we have some intuition as to how withdrawals may affect other depositors, observing this in real life is certainly complicated. As we have already found elsewhere in Volume 1, experiments can be very useful to separate different reasons and motives that influence individual behavior. This turns out to be especially useful in financial environments, given that at times of real crisis it is impossible to observe and separate out all the aspects that can affect the behavior of depositors. Do they all go to withdraw their deposits because they need the money, because of the information they receive, or because of what they see others do? To what extent do these explanations complement each other, in a similar way to which some factors (e.g. the existence of depositors with private information or deposit insurance) serve to mitigate withdrawals?[2]

In the following section we will define what a bank run is, by means of an example, and will look at the first theoretical explanations of the problem. In section 3 we study how experimental economics has analyzed the way that agents' behavior during the financial crises affected the availability of information on what was taking place. More concretely, we will see how this affects the *coordination* of the choices agents make to update their beliefs on what others are doing. Subsequently, we will see how agents update their ideas about the bank's situation or about the available investments, so that *herds* form and determine behavior. Finally, we will study what happens in the market for an asset or in a bank and how this affects the rest, so that financial problems become contagious. In all these cases, recent experimental studies have been able to determine the underlying mechanisms. Finally, in the conclusions we briefly revise the above-mentioned results and suggest open questions that can be explored in further research.

Why are there bank runs?

Though the concept of bank runs will be made clear as we advance through this section, for now we can identify it with the scene that opened this chapter: thousands of depositors arrive *en masse* to withdraw the money that they have deposited in their bank. The question is, what triggers such behavior by depositors? Why do they all come to the bank at the same time to withdraw their money?

The literature gives two polar explanations. The first one is related to the health and functioning of the economy, of the bank system and of the bank itself. This is known as bank runs due to problems of "economic fundamentals." The second way to explain bank runs is associated with coordination problems between the depositors. Below we introduce a simple illustration that should help the reader to identify both explanations.

Coordination problems and problems of fundamentals: a simple example

Let's imagine that you live on an island and that some time ago you invested, together with four friends, a certain amount of money into a company. Each friend contributed £100. Thus, the company's initial capital was £500. Recently, the company has decided to carry out a project of great profitability. The idea consists of investing £250 in the purchase of cattle in nearby islands, then selling the meat in their island, which has very poor meat production. To this end, you and your friends chartered a boat a few days ago, for £50. It is known that when the ship returns, the shipment of meat will be sold without difficulty on the island, up to quintupling its value. The profit from selling the meat will be, therefore, £1000 (250 × 5 – 250).

After chartering the ship and acquiring the meat, the company has spent £300, so it still has £200 to finance the return of the shipment, which would cost £50 (the same price each way). You and your friends know that once the ship returns to the island, each one will not only receive his initial investment but also the part corresponding from selling the meat. In total, this implies that you could receive £230. This amount corresponds to your share of net profits (a fifth of £1,000), after adding your share of the money that has not been invested in financing the expedition (a fifth of £150).

To receive your earnings, you and your friends only need to wait for the ship to return with the shipment. However, you signed a contract with the company against unforeseen events, which gives you the possibility of coming to the company and asking for the immediate return of your money. In gratitude for having participated, the company will give you your initial £100, and also a small amount equal to £10 as interest.

In these circumstances both you and your friends have two options:

- You can wait together with all your friends and receive £230 when the ship returns;
- You can go immediately to the company and ask for the reimbursement of your £110.

In this context, if a person decides to withdraw his funds, he will receive his £110, and those who wait will obtain a profit of £260.[3]

But what would happen if two of them go and withdraw their investments before the ship returns with the shipment? The company has promised £110 to each of them, but it has only £200.[4] In such a situation, the company can give £110 to the first person who comes to withdraw his money and £90 to the second; or it can return £100 to each.[5] The only certain thing is that it cannot pay £110 to each of them and that, regardless of what the solution is, it would have important consequences upon those who have chosen not to withdraw their money: since the company will have to face two immediate payments, it will not be able to finance the return of the shipment. Thus, the purchased meat will rot in the nearby islands without being sold, and all those who have chosen to wait for the ship would lose their investment and not receive anything.

Though it may seem strange, the functioning of the banking system has much in common with the simple example we have just discussed. The banks invest their money in projects that turn out to be profitable for their investors, but the investors must wait for a certain period of time until the assets mature. What happens if many investors demand their money in the short term? These projects will not be carried out. Therefore, banks face a problem between profitability in the long term and the demand for liquidity in the short term. Even if they are healthy, they can end up suffering massive withdrawals of money, without there being what are known as problems of fundamentals. Such problems occur when the doubts that cause massive withdrawals are not doubts about the actions of others, but doubts about the viability of a business.

For example, imagine that you think that the ship can sink, that a storm is approaching or that pirates can steal the shipment. Though these possibilities are unlikely, would you decide to withdraw your money? And if you see that a friend has withdrawn his money, what would you do? Would you think that your friend has some information about the shipment and that he has probably withdrawn his money because there are problems with it? All these situations would suppose problems with the capacity the company has to make the payments, for it could end up in a situation of insolvency. In our example, nevertheless, there is no uncertainty as to the profitability of the project. It is known that all the depositors are going to receive £230 if they decide to wait. The problem arises when some investor thinks that the others are going to withdraw their money, which induces him to fear for the profitability of his project. Just as in certain contexts, withdrawing can result in one receiving more than if choosing to wait, the depositors can end up coming to the bank for their deposits. As contained in the spirit of the phrase declared by J. P. Morgan, referring to the financial crisis: "If the people would only leave their money in the banks instead of withdrawing it . . . everything would work out all right." (See Bankers Calm; Sky Clearing. New York Times, October 26, 1907). In this context, what is known as bank runs due to coordination problems arise.

Coordination problems and problems of fundamentals are the main explanations of why bank runs happen, even though it is true that these two explanations can also be combined. Next, we focus the discussion on the problem of bank runs to introduce these two explanations in more detail and the main contributions made to the literature in this area.

Bank runs due to problems of fundamentals

The emergence of economic problems in a country is directly linked to the solvency and liquidity of the banking system. In consequence, depositors can alter their beliefs on the solvency and liquidity of the banking system when facing a worsening of the level of GDP of an economy, a significant increase in the level of unemployment, an important decrease in the rate of growth, or a sudden decrease in the valuation of companies' managers. Such instability or distrust can result in depositors coming to the bank to withdraw their deposits. Thus arises what is known as bank runs caused by problems of fundamentals: panic caused by problems related to the *health* of the economy or of the banking system.

The works of Gorton (1988) or Calomiris and Mason (2003) – among many other studies – are useful for a better understanding of what can provoke this type of bank run, and how one can react to them. Gorton (1988), for instance, supports quantitatively the hypothesis that the economic cycle explains the appearance of bank runs. Calomiris and Mason (2003) complement these findings by empirically studying how certain specific attributes of the banking system (e.g. geographical fragmentation) or the types of shocks (at local or national level) could have affected the probability that the American banks would fail during the years of the Great Depression (1930–1933). In their study, Calomiris and Mason analyze in addition which bank characteristics could have made them more inclined to fail, and discuss the determinants of contagion at the local level (a question we will return to at the end of this chapter). Their results show that the fundamentals explain well why the banks failed, especially during the first part of the Great Depression.

In real life, there is also evidence supporting the idea that bank runs are due to problems of fundamentals. We can draw attention to the problems that took place recently in Bankia (the third biggest Spanish bank by number of deposits), which suffered massive withdrawals when it became public knowledge that there were some accounting failings, or the massive withdrawal of deposits at the beginning of 2015 in Greece, after the change of government. These examples make it evident that depositors are concerned with the health of the economy and of financial institutions, and they can rapidly react to such instability or bad news by withdrawing their deposits.

Though the works of Calomiris, Mason and Gorton represent very important contributions in this setting, their macroeconomic foundation separates them from the approach of this book, which is centered on experimental evidence.[6] In this regard, Klos and Sträter (2013) have conducted experiments that study how different beliefs on the condition of the economy can affect the withdrawals of deposits. In their model, the depositors have a private and imprecise signal of the quality of the bank, following the theoretical line of global games by Morris and Shin (2003) and Goldstein and Pauzner (2005).[7]

Klos and Sträter claim that the subjects tended to follow a threshold strategy, choosing to withdraw or to wait depending on the type of signal they obtain, and on whether such a signal is above or below a certain value. In addition, they fit their evidence to a k-level model, based on evidence that the depositors have bounded rationality at the moment of deciding. The experimental evidence in problems of contagion, which we will discuss below (e.g. Trevino 2013), supports the idea that the better the signal the agents receive on the fundamentals of their bank, the fewer the number of withdrawals taking place.

Bank runs due to coordination problems

Before we study bank runs due to coordination problems, it is necessary to ask if these problems are important or not. As we have already mentioned, Calomiris and Mason find that the fundamentals can explain why some banks failed during the Great Depression. The authors, nevertheless, have problems in explaining a considerable part of the problems the banks had during the last stage of the Great Depression. Ennis (2003) cites some examples that have occurred in history and argues that although the deterioration in the fundamentals could have been the most relevant factor, the explanation of bank runs due to coordination motives cannot be discarded as a possible explanation, which makes both of them seem relevant. More recent empirical analyses (see Davison and Ramirez, 2014; De Graeve and Karas, 2014) suggest that banks with weaker fundamentals are more inclined to suffer bank runs, but there is also evidence that banks with good fundamentals could also experience sudden, large withdrawals. Therefore, it is suitable to propose another route to explain the withdrawals; bank runs due to coordination problems.

The depositors of a bank can choose to arrive *en masse* to withdraw their deposits, even in the absence of problems in the fundamentals, if they think that other depositors are going to do the same. During the panic at Northern Rock a depositor, when asked about her reasons for withdrawing her money, replied: "*It's not that I disbelieve Northern Rock, but everyone is worried and I don't want to be the last one in the queue. If everyone else does it, it becomes the right thing to do.*"[8]

The pioneering work of Diamond and Dybvig (1983) models this decision to withdraw deposits as a coordination problem between depositors who decide simultaneously, without knowing the decision of other depositors. In their model, Diamond and Dybvig show that there can be two types of equilibria. In one of them, the depositors (as would happen in the example of the company that we discussed above) will wait and leave their money deposited in the bank, with the expectation of receiving a greater profit in the future. In the other equilibrium, however, a bank run will happen: the depositors will be afraid that other depositors would come to the bank to withdraw their money, so they will end up coming to withdraw as well. This happens because the model assumes that those who leave their money deposited will not receive anything if the bank ends up without funds; thus it is always better to withdraw if others do it than to deviate and wait alone.

In the model of Diamond and Dybvig, it is assumed that there are three periods: the first period in which money is invested, the second in which it is decided whether to wait or to withdraw, and the third in which those depositors who have waited receive their earnings (those who withdraw their money receive their earnings immediately). One of the major contributions of the model, which supposes a fundamental difference with classic coordination problems that the reader has seen in chapter 4 of Vol. 1, lies in the introduction of two types of depositors, so-called patient and impatient. What makes the difference between patient and impatient depositors is their preferences, given that the impatient ones will receive utility only if consuming in the first period, whereas the patient ones value consumption in both periods.

In the model, there is a production technology that transforms any unit initially invested in the first period and in R> 1 units in the second period, which makes the socially optimal solution the one in which only the impatient ones withdraw in the first period, and the patient ones wait to receive their (greater) earnings in the second period.

Multiple theoretical works have been developed after Diamond and Dybvig, showing how bank runs can take place in different environments. The basic experimental model to study bank runs à la Diamond and Dybvig consists of a coordination game (in the style of a stag hunt game) where there are players interested in coordinating (patient) and others not interested in doing so (impatient). For instance, Arifovic *et al.* (2013) propose a model following the above-mentioned tradition, where the probability that each of the equilibria occurs will depend on how complicated it is to coordinate. The above-mentioned *difficulty to coordinate* depends on the proportion of patient depositors who must wait, for it to be beneficial to wait.[9] By means of experimental evidence and simulations, Arifovic *et al.* (2013) demonstrate that this parameter determines whether either one or the other equilibria would arise, even when the fundamentals of the economy are kept constant. In other words, the more depositors are necessary for the payment in the last period to be greater than the payment for withdrawing immediately, the more bank runs will take place.

Bank runs and information about the behavior of other depositors

The essence of the model of Diamond and Dybvig (1983) constitutes the fundamental basis on which most of the later articles in the literature on bank runs are built. Many researchers have used the model to relax some of its assumptions, or to study for example what happens if the bank incurs the cost of liquidation if it leaves the project where it invested the money initially. Experimental economics has turned out to be a useful way of shedding light on some aspects that are not easy to model: such as what happens when depositors, instead of deciding simultaneously what they want to do with their money, make decisions sequentially.

Most models of bank runs have traditionally chosen an approach to the problem in which agents *simultaneously* decide. However, the descriptions of bank runs (Sprague, 1910; Wicker, 2000 and Northern Rock's case) and some empirical studies (e.g., Starr and Yilmaz, 2007) suggest that many depositors have information about what other depositors have chosen to do, and can react to this information at the moment of making their own choices (Iyer and Puri, 2012; Kelly and Ó Gráda, 2000). The possibility of observing actions is obtained from many sources: the news of how other depositors are behaving, personal relations with friends who tell us what they have done with their deposits. To adequately discern how the behavior of the depositors is affected by the information received, experiments become essential, given that in real life it is very complicated (if not impossible) to know if the withdrawals are, or are not, affected by the type of information that depositors possess.

In their experimental study, Garratt and Keister (2009) form groups of five subjects, who have the opportunity to withdraw their deposits immediately or to wait. As in our example of the ship, the depositors maximize their earnings if they choose to wait (in their experiment, every subject gets $1.50 for every $1 deposited) but it is also possible to withdraw their money immediately. The bank is prepared to absorb two withdrawals (paying $1 to every depositor who withdraws immediately, and $1.50 to those who wait). If three depositors withdraw their money, they will receive $1 but those who wait will end up without money, and will not get anything. If there are more than three withdrawals, the bank has to liquidate its assets to face these demands: it would pay $0.60 to all those who have withdrawn; once more, those who wait do not receive anything. By slightly changing these payments, Garratt and Keister (2009) compare the case where the depositors have only one option to withdraw, with another in which they receive up to three options. In the second case, the subjects know how many persons in their bank have come to withdraw before the opportunity to withdraw their deposits is offered to them. The authors show that information about what has happened can increase the probability of withdrawals and, therefore, the occurrence of bank runs. In addition, Garratt and Keister consider a treatment in which some subjects are forced to withdraw, so that when others observe what has happened in their bank they see withdrawals but do not know if these have been forced or not. This characteristic in their model rescues the idea of the model of Diamond and Dybvig (1983) where there are impatient depositors who need their money urgently. In addition, the authors consider *aggregate uncertainty* by not informing subjects about the number of group participants that will be forced to withdraw. Their findings demonstrate that this type of uncertainty is also key to explaining bank runs due to coordination problems, given the greater probability of depositors withdrawing when the aggregate liquidity demand is unknown (when it is known, bank runs are in fact rare).

With the aim of better understanding how depositors react to the information they receive, Kiss, Rodriguez-Lara and Rosa-Garcia (2014a) conduct an experiment where the players form banks in triads. Two of the depositors of the bank are participants in the experiment, whereas the third depositor is simulated by the computer and has been programed to always withdraw his money from the bank. The subjects know of the existence of the computer so there is no aggregate uncertainty on the number of forced withdrawals. In the experiment, decisions are sequential (each of three depositors knows his position in the line) and it is possible to observe what has happened depending on the assigned information network. In some cases the decisions are simultaneous; in others they are sequential and what has happened in the bank is known. In yet others there is partial information (for example, the third depositor knows what the first one did, but the second one does not know; or, the second one knows that the third one will see his choice, but he does not know if the first one withdrew his money). Following the tradition of Diamond and Dybvig (1983) the subjects also know that both of them want to maximize their payoffs, but a depositor at the beginning of the line may be tempted to withdraw his money if he thinks that the subject choosing after him, and whose decision he cannot observe, will do the same. The theoretical prediction in Kiss *et al.* (2014a) demonstrates that if decisions are simultaneous (and every depositor decides without knowing what has happened in his bank), there is equilibrium multiplicity, as in the model of Diamond and Dybvig. However, if the depositors know that they are being observed and can observe what others have done, there is only one equilibrium without bank runs. The reasoning behind it is simple. Any patient depositor who acts first is going to leave his money deposited to induce the other patient depositor to act in the same way. Following this argument, if the first depositor decides to withdraw, any patient depositor acting later should infer that the withdrawal is due to the impatient player (simulated by the computer); thus, he should wait, even when observing a withdrawal. The subjects' behavior, none-theless, does not seem to correspond with this theoretical prediction, as shown in Figure 5.1. The evidence demonstrates that patient subjects react by withdrawing on having observed withdrawals, even if this was not an equilibrium.

The relevance of this article comes from having demonstrated not only that subjects can rush to withdraw their money after observing what others have done, but also that they do it out of panic, even in situations where the theory predicts that no bank run should happen at all. These results are in line with the experimental evidence of Schotter and Yorulmazer (2009), who already mentioned the import-ance that sequential decision-making has on behavior during bank runs. In their case, nevertheless, both environments (sequential and simultaneous) produce the same prediction; something that does

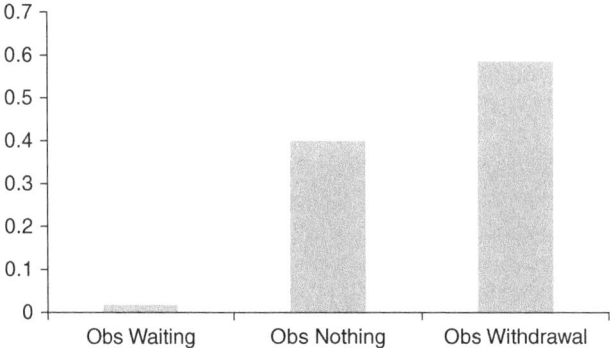

Figure 5.1 Likewood of withdrawal after observing the behaviour of other depositers. Graph based on Kiss *et al.* (2014a)

Note: "Obs" means observing the behaviour of other depositors. Waiting (Nothing) [withdrawal] means that the other are waiting (doing nothing) [withdrawing money].

not happen in the model of Kiss *et al.* (2014a), where the possibility of observing what other depositors have decided should produce a unique equilibrium without bank runs.

Other factors influencing the withdrawal of deposits

Already we have seen that the difficulty of coordinating, the presence of impatient depositors, or the possibility of observing actions can affect, among many other factors, the withdrawals of deposits. With the aim of studying what factors affect the dynamics and severity of bank runs, Schotter and Yorulmazer (2009) consider different treatments, where they vary not only the way in which the decisions are made (simultaneous or sequential), but also the type of information subjects have on the "quality" of the bank that determines, in the end, how difficult it is to coordinate. The depositors know that the money they have invested is going to provide them a given profitability, and that the bank is prepared to assume a certain number of withdrawals, depending on its quality. The experiment demonstrates how uncertainty as to the bank's quality, and how the presence of insiders (subjects with information about the profitability of the deposits) can influence withdrawals of funds (concretely, the insiders make bank runs less frequent and withdrawals are delayed in those settings where the depositors have various chances to withdraw their money). In addition, Schotter and Yorulmazer study how deposit insurance can affect the results, and find that even partial deposit insurance significantly reduces the rate of withdrawals. In this respect, Kiss *et al.* (2012) re-emphasize the importance of deposit insurances and the possibility of observing past actions, by arguing that both aspects are imperfect substitutes (i.e., the optimal deposit insurance should depend on how the information network is: that is, on the type of information that depositors may have).[10] Davis and Reilly (mimeo) study how the behavior of depositors should be influenced by the attitude the authority has (firm or indulgent) at the moment of suspending the payment to the depositors and renegotiating the terms of the agreements. Closely in line with the above-mentioned studies, the possibility of observing past actions and the attitude adopted by the authority will influence the probability of bank runs. For example, observing previous decisions will debilitate the positive effect that a firm attitude has, reducing the occurrence of bank runs in the case of indulgent attitudes.

An interesting contribution that tries to explain why withdrawals happen is that of Dijk (2014) who studies whether situations of fear, in and of themselves, are really capable of producing bank runs. To do so, before the subjects face a game of bank runs, they are induced into different psychological situations. Dijk observes that in the treatment with induced fear (where subjects must remember and describe with some detail the event in their lives in which they experienced most fear) the incidence of withdrawals is significantly higher than in other treatments (in those where they are asked to describe their happiest event or, simply, any random event). This opens the door to the interpretation of bank runs due to fear. The question is: is bad news spread by the mass media during times of crisis a way of inducing fear and provoking more panic? We leave reflection on this question to the reader.

Herd behavior and financial markets

The possibility of observing actions can affect depositors but does not need to be related only to coordination problems. The sequencing of actions may also relate to problems of fundamentals, as demonstrated by the theoretical analysis of Gu (2011). Gu analyzes how observing what the other depositors do affects the *beliefs* of the depositors following concerning the state of their bank. In this context, it shows how the process of *herds* can be generated.

To observe what others do and to update our impression of an issue is something that happens in many areas of our day-to-day lives. Before buying a new television or changing our mobile phone, for example, we are in the habit of looking at the list of the best-selling products. "If it is the most sold, there must be a reason for it," we tend to think. And it is curious but this is not only the reasoning we use when we want to buy an object for which we do not have information (how many times have we gone to buy a gift for a friend and asked the salesman what the best-selling product is?). It also happens when we have certain information (though probably incomplete) about the product that we want to buy or the market that we are in. In general, we tend to assume that the *majority* possesses certain information that can be useful at the moment of deciding.

A classic example is that of a restaurant in an unknown town. Let's imagine that we are visiting a tourist city and a friend has recommended a restaurant to us for dinner. Though our taste is not totally like that of our friend, we consider his opinion to be useful information so we decide to go to the restaurant. But when we get to the door we see that the place is practically empty, though there is another place just to the side replete with people. On having looked at the menu, we see that the menu and the prices are very similar in both restaurants. How many of us would leave the information we have aside (the advice of our friend) to follow what we see the majority doing? How much weight should the decision of the majority hold when we offset it against our private information?

These are the questions considered by the literature on *herd behavior*, referring to the tendency agents have to ignore their private information to follow the majority. In financial literature it has been suggested several times that herd behavior might serve to explain the excessive volatility of prices (excess price volatility) in the financial markets and the fragility of financial systems. Seeing many people buy an asset (stocks of a company) can make a person follow the herd and also buy one, even though that person has unfavorable information about the asset.

Herd behavior and types of informational cascades

The theory of herd behavior begins with the seminal studies of Abhijit Banerjee (1992) and Sushil Bikhchandani *et al.* (1992). In their models, the agents have private information about a good that they are interested in acquiring and must make their decisions in a sequential manner after observing what the other agents have done. The authors demonstrate that in certain environments, after a finite number of decisions, the agents will end up making the same decisions that the predecessors made, ignoring their private information. This situation is known as *informational cascades*, and although on occasion this could lead the agents to making an incorrect decision (for example, to go to an inferior restaurant, to buy an article of low quality or to invest in an asset that does not have value), they turn out to be rational decisions from the individual's point of view, provided that the information given by the herd can be evaluated as more valuable by an individual than his own private information.

In an informational cascade, the agent decides to ignore his private information and to follow the pattern established by the market. At first, we can distinguish three types of cascades depending on the behavior of the agent:

1. *Behavior of the herd* (herding): The agents ignore their private information and follow the pattern established by the market. For example, after observing many purchases, an agent decides also to buy the asset, regardless of his private information.
2. *Non-conformity* (contrarianism): In this case, the agents decide to ignore their private information and act in opposition to the pattern established by the market. For example, after observing many purchases, the agent decides to sell, and does so independently of his private information.
3. *Non-trade cascades:* These occur when the agents decide to abstain from the exchange, so they decide to neither buy nor sell the asset, regardless of their private information.

As we will see later, a key factor in determining if herds exist or not is the existence of an exchange price. In the example that we have provided of the restaurant, the exchange price is fixed, as the number of people that decide to go would not influence the price of the menu. In financial markets, however, the exchange price is affected by the decision of agents to buy or sell the financial assets. In both contexts, the agent's decision to follow his signal or to behave in accordance with the herd will depend on the measure known as trade imbalance. This refers to the difference between the number of purchases and sales that have taken place before an agent makes a decision. In this way, when we speak about a trade imbalance in a market of 2, we want to say that there were two more purchases than sales prior to the decision of the agent (in the same way, an imbalance of -2 implies that the number of sales is great than the number of purchases by two).

Informational cascade models in financial markets: the importance of exchange prices

To better understand the formation of informational cascades, let's consider the following model based on the work of Cipriani and Guarino (2005). Let's suppose that an asset exists that agents can exchange with a market maker. The fundamental value of the assets (V) is a random variable, with two equally probable values: zero or 100. The agents make their decisions sequentially in an exogenously determined order, after finding out the exchange price. The agents also have a signal regarding the assets. This (private) signal is informative so that if the value of the assets is V = 100, the probability that the agent has the signal that indicates a value of 100 is 70% (if the value of the assets is V = 0, there is a 70% probability that the agent has received the signal that the value is zero). The agents begin with an initial amount of money K> 0, and the final payment will depend on the difference between the actual value of the assets (V) and the price at which the asset has been bought or sold (P_t). Therefore, if an agent decides to buy it will receive $V - P_t + K$, obtaining $P_t - V + K$ if it decides to sell (if his decision is to do nothing, then the agent will receive K).

Agents not only have their private signal and price, but they also have information on the history of exchanges and of prices. The pioneer studies originating the literature of herd behavior (Banerjee 1992, Bikhchandani *et al.* 1992) did not have a price mechanism reflecting previous decisions, thus price P_t was not affected by the decisions of the depositors (this is what happens in case of the restaurants). To capture this idea, Cipriani and Guarino (2005) study the case of *fixed prices*.

Given the assets can take both values (zero and 100) with the same probability, the market maker fixes a price equal to the expected value ($P_t = 50$) in a context of fixed prices. The price does not change during the experiment, regardless of the decisions the agents make. Based on the logic derived from Bikhchandani *et al.* (1992) it is possible to demonstrate that in such an environment, an informational cascade can result after a trade imbalance, greater or equal to two or lower or equal to –2, is formed. To understand this result, imagine that the third agent in deciding observes that both previous agents have bought the assets (that is, there is a trade imbalance of two). Also suppose that the third agent has a signal telling him that the value of the assets is zero. Should the agent follow his private signal and sell the assets for a price $P_t = 50$ or should he buy the assets, though his signal indicates that the price is zero? Notice that, after observing two purchases, the third agent will deduce that the signals corresponding to both previous agents were of 100. Bearing in mind these inferred signals and his own private information, the agent might apply Bayes' rule to deduce that the asset's expected value is 70.[11] Provided that the exchange price is 50, the third agent should ignore his private signal and buy, thus beginning a cascade because the next agent (though he has a negative signal) will see that there were three previous purchases, and will end up buying as well.

A characteristic of financial markets is that agents' decisions reflect on the asset's price. If many of them decide to purchase (to sell) assets, its price raises (falls). This idea appears in case of *flexible prices*, in which the market maker is not forced to keep the same price in all periods, but he can update the price according to Bayes' rule, bearing in mind what the decisions made by the agents have been. Following the idea of Avery and Zemsky (1998) it can be shown that when the market maker establishes the price according to Bayes' rule, agents will follow their private signal, and therefore informational cascades cannot take place. The intuition behind this result is quite evident. Since the market maker updates the market price using Bayes' rule, *the price contains all the information from the previous transactions.* Therefore, a rational agent should act following his private signal, which has additional information on the value of the asset. Given that this information is unknown to the market maker, the agent can capitalize on this informational advantage to his benefit.

If we return to the previous example, in which the third agent observes a trade imbalance of two, and we use Bayes' rule, we can calculate that the market price offered by the market maker will be 84.48 (not 50, as in the fixed price case). Since the agent has a private signal telling him that the value of the assets is zero, he will use this additional information (applying Bayes' rule) to calculate that the asset's expected value is 70 and consequently that he must sell the asset.

In Figure 5.2, we represent the Bayesian updates. We begin at 50, the asset's unconditional price. In the first period, after receiving a positive (negative) signal, the updated price using Bayes' rule will be 70 (30), thus the first agent should buy if the signal is good and sell if it is bad. This decision directly reveals the first agent's signal. In the second period, if the trader receives a different signal from the one received by the first agent, then the update gives a price equal to 50. If an identical signal to that of the first agent is received (which, as above-mentioned, can be inferred from the decision he has made), then the agent can be certain that the price is either 100 or zero. Therefore, we move towards the extremes. One could see that prices

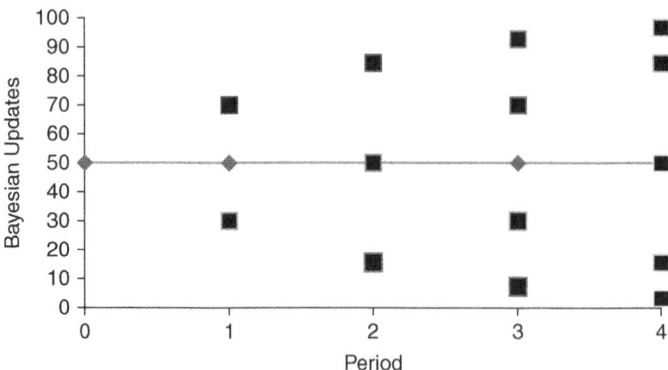

Figure 5.2   Bayesian updates in markets with fixed and flexible prices

Note: Diamonds indicate the asset's unconditional price. Squares represent the update price after receiving the good/bad signal.

are located on a grid and opposite signals are canceled. What really matters in these cases is the trade imbalance: that is, how many more purchases than sales have taken place.

In the fixed price case, if at some moment the number of purchases is higher than the number of sales by two, in spite of receiving a negative signal, the price updated à la Bayes will be higher than the price established by the market maker (50). This motivates the agent to buy, affecting by his behavior all those who act after him. Using the same argument, a trade imbalance of –2 makes agents, who still have not made their choice, sell – independently of their private information. When the price is flexible, the market price after two purchases is 84.48. If the third agent receives a positive signal, he will buy because by applying Bayes' rule with his private information it will lead him to calculate an expected value of 92.7 for the assets (as above-mentioned, the third agent will sell if his signal is negative because his updated value will be 70, which is lower than the market price offered by the market creator, 84.48).

Experimental evidence of herds in financial markets: fixed and flexible prices

Empirically, it is difficult to test herd models, because there is no data on the private information of agents. In addition, it is difficult to know if agents react differently depending on whether the exchanged asset's price is fixed or flexible. This can be done in the laboratory, where subjects can receive private information on the value of the asset and observe the history of exchanges.

Cipriani and Guarino (2005) used 216 subjects to test differences in herd behavior depending on prices. Their experiment consisted of ten rounds, in each of which 12 subjects had to make their choices sequentially. Before every round, an experimenter would toss a coin to determine whether the value of the asset was going to be zero or 100. The participants did not know the result of the coin toss, but they received an informative signal. There were two bags in the experiment. One contained 30 blue tokens and 70 white tokens; the other contained the opposite. The bags were identical. Each subject, when asked to make his choice, had to draw a token from the bags. If the coin toss resulted in heads (tails), the participant will draw a token from the first (second) bag, so that the color of the token was an informative signal on the asset's value in that round.

Each participant, after privately observing the color of the token, was to return the token to the sack and to announce out loud if he wanted to buy, to sell or to not do anything with the assets. The price of exchange was determined by the asset's price, fixed on the board. Also, the decisions and prices that had taken place so far were registered on the board, so that participants were not only receiving their private signal, but they were also obtaining information about the history of transactions and prices. The experimenter updated these prices depending on the treatment.[12] Remember that in the fixed price treatment the price is constant and equal to 50, while in the flexible price treatment it changes according to the decisions made. In Table 5.1, adapted from Cipriani and Guarino (2005), we see a summary of the participants' behavior in the different treatments.[13]

Table 5.1 Table extracted from Cipriani and Guarino (2005). The relevant periods are those where it is possible for informational cascades to be produced, for the trade imbalance is at least two (or at most −2) and subjects receive a negative (positive) signal.

The theory predicts informational cascades when the price is fixed, as long as the trade imbalance is at least two (in absolute value). Throughout the experiment there were 58 opportunities for potential informational cascades to emerge. In 52% of the cases, informational cascades occurred, where subjects stopped following their private

Table 5.1   Table extracted from Cipriani and Guarino (2005)

	Fixed Price	Flexible Price
Act against the private signal (herd)	52%	12%
Do nothing	26%	42%
Act according to the private signal	22%	46%
Relevant periods	58	66

signal to act in line with the behavior of the majority. In this treatment, subjects decided to do nothing in 26% of the cases, whereas 22% of the times they followed their private signal. As illustrated in the table, things were very different when the price was flexible. In this treatment, there was 66 periods in total where the trade imbalance was at least two and subjects received an opposite signal. An informational cascade was formed in only 12% of the cases, with subjects ignoring their own signal. In 42% of the cases, subjects decided to do nothing and in 46% of the cases they followed their private signal.

These results are in line with the theoretical prediction, according to which we should not observe herds with flexible prices but only with fixed prices.[14] In general, this directly highlights the importance that an asset's price has – to aggregate or transmit the information different subjects have on the asset's value.

Non-conformism and no-trade cascades

A strange phenomenon observed by Cipriani and Guarino is that many subjects decided to do nothing although in theory this is never optimal. In the treatment with flexible price, in addition, the frequency of not doing anything increased with the absolute value of the trade imbalance. The higher the trade imbalance, the closest the price was to the extreme values (zero or 100), so that not doing anything can be understood as a response to the great risk implied by making the wrong choice. That is, when the market price is close to 100, the subjects were afraid that such a price was probably given by mistake and that the real value of the asset was zero.

Another phenomenon observed in the experiment was that of decisions "against the market," which has been denoted as contrarian behavior. In the flexible price treatment there were 132 occasions where subjects could act as non-conformists. They behaved this way in 19% of the cases, 18% decided not to do anything and 63% followed their private signal. In cases of contrarian behavior the market was not able to correctly aggregate information. In the fixed price treatment, subjects never adopted the contrarian behavior. These results go in line with the observations in Drehman *et al.* (2005). In their Internet experiment the flexible price made it unlikely for herd behavior to appear, though there was evidence of contrarian behavior in these types of markets.

Financial market professionals: external validity

When speaking about experiments in economics the problem of external validity often arises. Will we be able to extrapolate what we observe in the laboratory to the real world, where the environment is less controlled and generally much more complex?

Though there is evidence suggesting that experimental subjects do not differ much from the rest of subjects in the population (Exadakytlos *et al.* 2013), it is always good to contrast the evidence obtained in the laboratory against the decisions other subjects make, especially if the aim is to know how experts in financial markets behave. With this idea in mind, Drehman *et al.* (2005) study the behavior of 267 consultants from McKinsey, the company, and conclude that their decisions were not significantly different from those made by other subjects from all kinds of backgrounds, in a very similar experiment to that of Cipriani and Guarino (2005). In the environment of cascade formation, Alevy *et al.* (2007) study if there are behavioral differences between students and professionals of the *Chicago Board of Trade*. They find that the decisions students make are slightly more consistent with the theory (that is, they do not contradict what is predicted by a Bayesian Nash equilibrium). However, the rate of cascades formed is not different between students and professionals, though professionals use their private information much more than students do, which leads them to *fewer wrong cascades*.

Cipriani and Guarino (2009) analyze the behavior of 32 professionals employed in financial institutions in London, to see if the results we have just discussed were robust to this manipulation.[15] Their results demonstrate that the professionals' behavior was not very different from that of students. Abstaining from trading kept arising as a relevant behavior and though the professionals had a low trend towards herd behavior, they tended to be less conformist than students.

The connected economy: contagion problems

Up to this point we have studied what happens between participants in a specific market, or the depositors of a specific financial institution. But for instance, what would happen if Greece decides to leave the euro? Would it affect the rest of the countries in the Eurozone? There is no doubt that the interdependency between agents and economic markets is one of the most relevant characteristics of contemporary economies. One of the crucial aspects of increasing globalization is the connection established between individuals, companies and countries via constant flows of trade and information. This causes phenomena to be rapidly transmitted from some individuals to others, from some zones to others. During the recent Great Recession we have seen how problems in a certain country were rapidly spreading to neighboring countries, such as the bankruptcy of certain companies being transferred to different parts of the production chain, and how financial problems in certain organizations made others, who were apparently robust, vulnerable. In fact, the problems that financial institutions suffered are regarded as one of the principal engines of the Great Recession. The initial trigger was the explosion of the sub-prime mortgages that, through toxic assets, unexpectedly flooded the financial system. The innumerable connections between banks caused problems to hop from one organization to another and one country to another.

The characteristics and problems generated by the different connections between economic agents have recently kept economists busy. Some studies (for example, Babus, 2014) have showed how certain connection structures between banks can have different effects when financial problems arise. In general, a greater degree of bank interconnection allows the banks to have more diverse financing and investments, although strong interconnections can also make the bank system more fragile with regard to aggregate impacts. In such cost–benefit analysis it is fundamental to clearly understand what advantages are provided by a certain network structure (Cabrales *et al.* 2014). Technically, this is a complex matter since the interactions between agents frequently lead to situations with multiple equilibria.

Studying what types of structures emerge and how different structures can affect outcomes is a question of some interest from the experimental point of view.

In their attempt to answer some of these questions, Corbae and Duffy (2008) study a setting where individuals first propose forming connections between them to, afterwards, play a coordination game with their connected neighbors. The formation of connections can be interpreted as the decision banks make when they want to minimize their risks. Confirming their theoretical prediction, the authors observe that bilateral structures are most frequent in the experiment. In their model, these connections are the most efficient and the ones that allow the changes from contagion to be minimized.

Contagion problems are also the aim of study in two recent works, where Brown, Trautmann and Vlahu (2014) (BTV hereafter) as well as Chakravarty, Fonseca and Kaplan (2014) (CFK), explore how the existing problems in a bank can affect the depositors of another bank. In both cases, the structure is given. The depositors of a bank observe what the depositors of another bank have decided, where both banks may or may not be linked. If I hear the news that a bank has suffered problems, will this lead me to thinking that my bank could also have these problems and to withdraw my deposits from it? Both studies find that such contagion takes place when the banks are linked. That is, when the situation of another bank (good or bad) can be indicative of the situation of my bank, observing the problems in the other bank will lead the depositors to withdraw their money. Though one could expect this to happen, CFK also find that even if the depositors are fully aware that both banks are completely independent, observing problems in the neighboring bank can provoke the depositors of the unrelated bank to also run for their deposits. Let's see in more detail how these studies are conducted.

BTV study pairs of banks, each formed by two depositors. In each bank, depositors participate in every period in a bank run game, such that if both depositors wait, they get a high payment. If both depositors withdraw they get a low payment; but if only one of them withdraws, he receives an intermediate payment and the depositor who waits does not receive anything (the bank ends up without funds). The banks can be in a good or a bad situation, characterized by a greater or a smaller payment in case both depositors choose to wait until the end. To study the possibilities of contagion, three experimental treatments were conducted. In the control *group*, the depositors of the second bank made their choices without knowing if their bank was in a good or a bad situation, and without knowing what had happened in the other bank. In the *treatment* with *link between banks*, the depositors of the last bank also ignored if their bank had good or bad fundamentals, though they decided what to do after observing what had happened in the first bank, knowing that the fundamentals of their bank were the same as those of the first bank.

Finally, in the treatment *with no link between banks*, the depositors of the last bank to decide ignored whether their bank had good or bad fundamentals and made their choices after observing what had happened in the first bank, knowing that the fundamentals of their bank were completely independent from those of the first bank. To elicit information about the beliefs of the depositors, BTV used a seven-point Likert scale, to ask each depositor how likely or unlikely it was for them that their bank had good fundamentals, and whether they thought that the other depositors in the bank were to withdraw their deposits or not. Through this, the authors were able to identify the mechanisms through which panic becomes contagious between banks.

The results of BTV show that when there is a link between the banks (they share fundamentals), the depositors, after observing withdrawals in the first bank, assign a higher probability both to the idea that the fundamentals of their bank must be bad, and to the expectation that the other depositor in the bank will withdraw his deposits. This shows that both routes of contagion, due to beliefs in the fundamentals and due to coordination, take place in linked banks. Another relevant finding is that when the banks were not linked, contagion was not observed, provided that the withdrawals and beliefs were not different from the control group.

CFK studied banks formed by ten individuals, who choose for 30 rounds whether to withdraw their money or to wait. In their model, half of the depositors were assigned to the impatient role, and they had a greater payment if they withdrew their money, whereas the other half were patient, and had a greater payment if a sufficient number of them decided to wait. The fundamentals of the bank could be good or bad. In the case where they were good, the payment for the patient depositors was higher if they waited even if the other two patient depositors were to withdraw (out of five). If they were bad ones, once a patient depositor withdrew his money, it was best for each participant to withdraw.

After the depositors in one of the banks had made their choices (they knew the condition of their bank), the depositors of the other bank were to decide, after observing what had happened in the first one. Though they were not informed about the real condition of their bank, they did know that the fundamentals were good in each round with 80% probability. To understand the cause of contagion, they conducted two treatments. In the first one, the fundamentals were linked, so that both banks were in the same situation (and this was known by all the depositors). In the treatment with independent fundamentals, the situation of each bank was independent.

CFK also find evidence for contagion. When it is observed that a bank run has happened in another bank, it increases the probability of withdrawals by the patient depositors. Though this happens to a greater extent when the fundamentals of both banks are connected, an increase in withdrawals is also observed when the depositors know that both banks are independent. This differs from the result found by BTV.

The sequential decision-making in the experiment of CFK allows them to also observe how bank run situations are extended over time. A relevant finding in this regard is that, though observing a bank run in the other bank increases the chances of a bank run by the depositors, observing that the bank run stops does not stop the panic the depositors feel. That is, the contagion occurs in one direction, provoking panic, but there is no such thing as "anti-panic contagion".

As a whole, these two works have been able to experimentally show that bank problems are contagious from some to others (it is not simply that the depositors respond to the same circumstances). A third recent study on contagions that complements the previous ones very well is that of Trevino (2013). In this work, Trevino explores the possible routes of contagion when it is not possible to accurately know what happens in another bank. In her model, she considers the situations of sovereign debt holders in a country that depending on its real condition can end up suspending its payments or not. She suggests that another interpretation is that of bank runs. In her case, the bank or country can be in a good condition, in a bad condition or in a regular condition. The condition of the bank will determine whether it is best for a depositor to wait or to withdraw his money (when the condition of the bank is good, it is better to wait; when it is bad, the best thing is to withdraw; and if the bank is regular, it is best to withdraw if the depositor withdraws and to wait if the depositor waits).

To study contagion, Trevino assumes there are two banks, whose fundamentals can be linked or not. There is an a priori idea on whether the fundamentals are good or bad, common for both. To the above-mentioned information, a private signal given to the depositors of each bank is added, about the condition of their bank. The richness of the model comes from the fact that after the depositors of the first bank make their choices, the depositors of the second bank receive imperfect information about what has been decided in the first bank, which can help them infer the condition of the first bank.

The experimental treatments used by Trevino to understand how contagions take place consist of varying how correlated (not at all, moderately, or perfectly) the fundamentals of both banks are, and how much the depositors of the second bank know (nothing, moderately, or perfectly) about the actions of the depositors of the first bank.

In line with the explanation of withdrawals due to fundamentals, Trevino finds that the depositors of the first bank withdraw less the better the signal they receive. With regard to contagion, Trevino seems to support the idea of CFK, finding that when withdrawals are observed in the other bank the probability for withdrawals increases, even when the individuals knew that the fundamentals from both banks were independent. An additional interesting discovery by Trevino is that this happens even when the depositors know that the information they receive on the bank run in the other bank need not be linked to their reality. That is, if they are informed of a bank run in the other bank, even when knowing that such information could be either true or false, this leads them to increase their withdrawals. We can link these findings to the ones on psychological panic described by Dijk (2014).

Conclusions

In this chapter, we have surveyed some of the most relevant experimental studies on financial crises, emphasizing bank runs and how different information availability affects the agents' behavior. The experimental studies have shown the existence of bank runs both due to problems of fundamentals and due to coordination problems. In addition, they have shed light on the existing dynamic mechanisms which had not been well addressed in the theoretical literature.

In fact, several studies have shown that when people can decide after knowing what others have chosen this increases the probability of withdrawing. This leads us to thinking that an important part of why bank runs happen depends on fundamentally dynamic components. Coordination failures between agents, which lead to situations of panic, are much more credible when individuals observe what others do or when their decisions are repeated. In addition, they increase in situations of fear, when nobody has additional information or when facing less firm monetary authorities.

Observing what others do also leads to updating one's beliefs on the viability of the bank or the financial asset in which the agents take part. The experimental analyses have shown how herd behavior can emerge, which lead agents to make their decisions in similar ways, in spite of possessing different private information. In this respect, it has been shown that in situations like those of financial deposits, where the expected payment is fixed, the emergence of herds is much more likely than in situations like those of other financial markets, in which the price of the asset is updated with the decisions of the agents.

Finally, we have analyzed recent studies on the contagion of financial crises. These experiments have shown that when bank runs arise in a certain bank or massive sales of debt take place in a certain country it affects the rest, so that the crisis is transmitted from some assets to others. One of the most relevant findings in these experiments is that the above-mentioned contagion happens when either banks or countries are in similar situations, but also when they are completely independent. That is, financial crises in an individual bank or asset increase the instability in the rest of the system, even in those banks or assets isolated from the one that triggered the problem.

The experimental literature on financial crises is a relatively recent field and there are still many open questions. Most studies until now have focused on bank runs, but there are other markets where massive withdrawals of funds take place that have not been experimentally studied, such as the case of investment funds, the interbank market or the repo market, among others. The study of sequential problems has revealed how influential the information agents receive is, but little is known about who starts these processes and how they develop. Finally, although some studies have revealed implications as to what measures to take, there is still no systematic study of the optimality of the different actions carried out by governments and central banks to limit these processes. All this opens up the study of financial crises as an important field in which there are still many questions on which to experiment.

Notes

1. A recent example in this global context that the reader can think about is the Greek case, where many depositors decided to withdraw their deposits out of the country at the beginning of 2015 after the arrival in power of Alexis Tsipras, leader of the Radical Left Coalition (SYRIZA). Such financial instability in different countries, in fact, motivates the study of Trevino (2013) that we will discuss below, analyzing how the holders of bonds of a country can decide to withdraw their money out of the country.
2. The interested reader can extend his or her vision of the financial crises to other recent reviews of the literature with a slightly different approach to ours: Heinemann (2012) surveys a wide catalog of financial problems, including bubbles and herds; Duffy (2014) surveys the first experiments on bank runs; and Dufwenberg (2014) surveys experiments on banking, including also bank runs.
3. Let's see why: because now there are only four people with whom to divide the net profits for the sale of meat, each would receive £250. In addition, the company has £40 after financing the return of the ship and paying the person who has withdrawn his money, thus it would get ten additional pounds.
4. Because the company had £500, this is the amount it has after having bought the meat (£250) and having financed the expedition (£50).
5. Notice that the difference between each case is that depositors can be served in the order they come (£110 for the first one and £90 for the second), or the company could wait until they all have claimed their money before satisfying their demand. In the literature, the first alternative is known as *sequential service constraint.*
6. The reader interested in knowing more on bank runs due to problems of the fundamentals can see the review of the literature in Gorton and Winton (2003).
7. The experiments we will discuss in detail when we speak about informational cascade models also propose beliefs on fundamentals, assuming that each depositor has a signal on the health of the bank which he can use when making his decision.
8. See http://www.bloomberg.com/apps/news?pid=newsarchive&sid=ax6oy 1MsuSXA&refer=home
9. In our example of the company we have not introduced patient and impatient types as this would complicate in excess the illustration; but it is easy to see that just one person should withdraw before the ship returns (what we have denoted period 1) so that it is worth waiting. Notice that, if two of them decide to withdraw their money from the company, they will receive part of their investments, whereas those who wait will not

receive anything. If only one person withdraws (and gets £110), it is still beneficial for the rest to wait (and get £230) instead of withdrawing.

10. These authors have also studied how the gender or the cognitive skills can influence subjects at the moment of withdrawing their deposits in such situations of panic (see Kiss et el. 2014b, 2015).

11. To observe this outcome we denote by (b, b, 0) the information set indicating the first and second trader have bought (that is the reason for the two b's) and the third has received a signal that the assets is worth zero. Therefore, the asset's expected value is

$$E(V \mid (b,b,0) = 100 * \Pr(V = 100 \mid (b,b,0))$$

$$= \frac{100 * \Pr((b,b,0) \mid V = 100) * \Pr(V = 100)}{100 * \Pr((b,b,0) \mid V = 100) * \Pr(V = 100) + 100 * \Pr((b,b,0) \mid V = 0) * \Pr(V = 0)} = 70$$

12. At the end of every round, the realized asset's value was revealed and each subject received his earnings, paid once the experiment had finished. In this experiment, subjects could not lose money because in each round they received 100 lire (the experimental currency) and the value of the asset was always between zero and 100.

13. To elaborate the table, only the last seven rounds of the experiment have been considered, and the relevant cases in which (theoretically) it is possible to observe an informational cascade. That is, we focus on cases where the trade imbalance was at least two (or at most -2) and the subjects received a negative (positive) signal.

14. In the experiment, there was a third treatment (called "*without history*") used as control to better understand the effect history has. This treatment was identical to the flexible price treatment, but subjects could not observe the decisions made previously. In this treatment, only 24% of the subjects observing a trade imbalance of at least two decided against their signal, indicating that it was not history what was behind the results in the flexible price treatment. In a last treatment (called of "endogenous price") the update of prices was not realized following Baye's rule, but it was endogenously realized by some participants in the experiment, who were designated to do so (and who were competing between them). The results of the latter treatment are very similar to those obtained in the flexible price treatment.

15. In addition, Cipriani and Guarino studied the importance of informational uncertainty in the formation of herds, by introducing *informed agents* (with private information about the asset) and *noise traders* (who did not receive this information). In another study, Cipriani and Guarino (2008) studied the effect of transaction costs on the process of information aggregation by introducing a Tobin tax in the flexible price model. In this variation of their model, if an agent decided to buy or to sell he had to pay a fixed amount (lump-sum).

6
Labor Market: Incentives, Wages and Contracts

Enrique Fatas and Antonio J. Morales

Introduction

Prendergast (1999) begins his widely known survey of incentives of the firm with the sentence *"incentives are the essence of the economy."* Hardly any economist would disagree with the idea that economic agents react to the incentives they face, even more so if a significant number of individuals have motivations beyond their own self interest and have social preferences, as discussed in chapters 6 and 7 in Vol. 1.

Due to the central place that incentive design in firms and organizations has in economic science, it is somewhat surprising to find that the excellent survey of the literature of Prendergast (1999) subsequently states that *"despite the numerous and definite arguments regarding their alleged importance, there has been limited empirical evaluation of the provision of incentives for employees."*

In recent years, experimental economics has aimed to contribute to alleviating this scarcity. In this and the next chapter we will review the main contributions to the analysis of incentives in organizations both when the problem is associated with the predominance of free-rider problems in team work (as discussed in chapter 7 of Vol. 1) as well as with relations between employees and employers in organizations, the so-called principal–agent problem.

We will begin with the problem of incentives in horizontal organizations, where employees interact between themselves, in the presence of institutions or rules of the game that regulate their conduct and reward. In this setting we will see different ways of controlling for the ill effects of opportunistic agents who may try to maximize their utility by choosing low levels of effort. In the second part of this chapter, we will present the problem that firms face as an employer (the principal) to obtain a minimum response from the employee (the agent). We will see how the theoretical analysis is considerably enriched when brought into the laboratory in a controlled environment.

The firm as a team

The problem of team production

To define the team production problem we will use a simple environment where the production function of the organization is, for simplicity, linear. There is a double advantage to considering this case:

- Team production depends directly on the aggregate effort of their members.
- There are identical consequences from the efforts of each employee (x) on the collective performance.

The second point means that employees are, therefore, homogeneous in their resources and abilities, and there are no complementarities between their individual efforts. In consequence, production is simply the sum: that is $\Sigma_{i=1}^{N} x_i$.

Similarly to real life, the firm is unable to directly observe the individual effort each employee makes. There are different reasons for this: a) it might be very costly to implement a personalized surveillance system; b) it may be easily observable without being verifiable by third parties (for instance, before a judge who can evaluate the reasons to fire an employee given his low effort level).

The profit each team member gets is an increasing function of the collective effort and decreasing in own effort. From the firm's point of view, the more effort the employees make the greater its production level, and the higher the wages its employees will receive. From the employees' perspective, making effort is always linked to a personal cost, which must be considered jointly with the wage earned when choosing an optimal effort level.

We can represent an employee's profit as follows (see an almost identical expression in Chapter 7):[1]

$$\pi_i - \pi_i^p + \pi_i^g = (e - x_i) + B \sum_{i=1}^{n} x_i, \quad 1 > B > \frac{1}{n}$$

Thus, the profit an employee gets is composed of a private gain π_i^p and a profit for group work (the firm or the team) π_i^g.

- The private profit captures the cost of effort, which we assume is unitary for simplicity (for each unit of individual resources e invested by making an effort in the firm x_i costs him one unit). Therefore,
  - e represents the employee's individual resources (for instance, his work shift, his abilities, skills, etc.),
  - x_i constitutes his individual effort level (for instance, the number of hours he really invests his maximum effort, in his work shift).
- The profit π_i^g he gets from his work depends on the effort of all other i employees, and it is a function of B, a parameter capturing the positive externality associated to performing a task in the group, and of N, the size of the group.

Let us try to understand the decision-making process an employee goes through. Dynamics in this model are interesting in spite of its extreme simplicity. Each employee compares the opportunity cost of exerting effort (the private gain negatively depends on the chosen effort level) against the profits associated with group productivity (group gains π_i^g positively depend on the collective effort).

Under known assumptions, rationality and common knowledge, it is not complicated to calculate the optimal choice. Given that $B<1$, and that group profits $\left(\pi_i^g\right)$ are acquired independently of the individual effort level, regardless of the effort level exerted by the rest of the team members, the best response is to always choose the minimum effort level $x_i = 0$. We can identify this strategy as the only Nash equilibrium (NE) of the game, applying the backward induction argument (see chapters 3 and 7 of Vol. 1).

From the firm's perspective this solution is not close to optimal. For the firm, the best outcome is that in which the effort all employees exert is maximal. In fact, the employees are the most interested in achieving a positive level of effort, because if all choose the minimum effort level their collective gains $\left(\pi_i^g\right)$ are zero. Given that the effort each team member exerts has a positive outcome for all its members, and that the marginal productivity of each unit of effort *(B)* is strictly greater than *1/N*, the solution that maximizes collective profits[2] consists of choosing a maximum effort level $x_i = e$. In such a situation, each team member's profit *(B·N·e)* is greater than if no one exerted any effort *(e)*.

In other words, the team production problem arises when it is impossible, or very costly, to monitor employees' individual effort levels, and there is a tension between the individual and the collective. This tension is translated into a struggle between the individual rationality from NE (not exerting any effort) and the collective rationality of putting maximal effort (which is not an individual's best response to the actions of others). Alchian and Demsetz (1972) were, probably, the first to identify this dilemma employees face: assuming the total costs of individual effort while only receiving 1/N of the collective profit.

These types of interactions have been studied in the laboratory both as team production problems and as voluntary contribution games (see chapter 7 of Vol. 1). As the reader may have observed, the team's profit satisfies the two conditions in a pure public good: it is not exclusive and is non-rival. The work by Ledyard (1995) extensively analyzes this experimental literature.

In these laboratory works we observe that results greatly differ from the strict theoretical prediction, but also they are notably different to the optimal solution linked with maximum level of effort. In addition, the results are closer to the NE as the game is repeated. Croson *et al.* (2005) is a good example of this dynamic. Average effort levels begin at around 50% of the individual's resources. As is common in this type of game, the observed effort levels decrease notably through the game towards the only NE, so that average effort levels are hardly above 10% of resources in the last period.

Punishments, rewards and competitive incentives

Naturally, a question that arises is how we can avoid the collapse in productivity in firms exposed to team production problems.

One natural response is to introduce a monitoring system to spot employees that choose a low effort level. The seminal work by Nalbantian and Schotter (1997) shows that systems based on random inspection and punishment of free-riders (those employees with an arbitrarily low effort level) only work when the probability of inspection is inordinately high.

Fatas, Morales and Úbeda (2010) analyze the introduction of random punishments. In their study, each employee is punished with the loss of their group profits $\left(\pi_i^g\right)$ with a probability that inversely depends on their team's performance R. In the extreme, if the team has an optimal performance, the individual probability of being punished is 0% while if the performance is nil the probability is 100%.

We incorporate punishments to the above-presented profit function to get a new payoff function $\left(\pi_i^s\right)$:

$$\pi_i^s \begin{cases} (e - x_i) + B \sum_{i=1}^{N} x_i & \text{with probability } R \\ (e - x_i) & \text{with probability } (1 - R). \end{cases}$$

The probability of being punished is identical for each team member; therefore, the firm does not need to search for individual effort levels, reducing its implementation cost. However, this mechanism is unable to detect the effort of those team members who do exert effort, as the probability of being punished is the same for all its members. It is not difficult to prove that the mechanism does not change the theoretical prediction of the game; through backward induction and assuming subjects are risk-neutral. That is, the NE predicts that subjects will exert no effort.

The outcome of this punishment mechanism is portrayed in Table 6.1, where it is observed that, independently of whether it is

Table 6.1 Productivity and penalties

	Mean effort (%)			
	Block 1		Block 2	
Period	*1*	*1 to 20*	*21*	*21 to 40*
No sanctions	38,62	27,07	37,00	20,38
With	54,06	50,40	48,98	38,35

applied from the beginning (rounds one to 20) or after a 20-round block without punishments (rounds 21 to 40), random punishments almost double the effort levels and, in some way, prevent team performance from declining.

The impossibility of discerning between different effort levels within a group imposes an important cost. By being punished in spite of working hard, the most productive employees react to such punishment by persistently and dramatically decreasing their effort levels. On the contrary, the least productive employees, although they exert more effort than the least productive employees in settings without punishments, do not significantly react to the punishments.

To reduce the negative effects on the most productive employee's efforts, recent literature has extensively analyzed the option of incorporating into the firm one of the adjustment mechanisms most liked by economists: the market. The idea was presented on theoretical grounds by, among others, Baker, Gibbons and Murphy (2001), and it argues in favor of incorporating competitive mechanisms that automatically allocate the best wages to those firm members who win the competition.

Nalbatian and Schotter (1997) found that distributing a small fixed prize between two teams of employees competing for it, as a function of their relative performance, was a powerful incentive that yielded better results than any other monitoring system or system of contracts based on objectives.

Croson, Fatas and Neugebauer (2006) analyze this type of competitive mechanism in the shape of a punishment system similar to those expressed in the previously presented profit functions. The team member with relatively lowest effort was punished and lost π_i^g The payoffs π_i^c associated with this competition are represented by the following expression:

$$\pi_i^c \begin{cases} (e - x_i) & \text{if } \max\{x_j\} > x_i = \min\{x_j\}; j = 1,\ldots,n \\ (e - x_i) + B\sum_{i=1}^{N} x_i \end{cases}$$

Otherwise this mechanism simply establishes that an individual with the lowest contribution to the group (if there is any other employee exerting a greater effort) is punished; this means that he does not receive the public good.

Table 6.2 Productivity and competition

	Mean effort (%)			
	Block 1		Block 2	
Period	*1*	*1 to 10*	*11*	*11 to 20*
No sanctions	40	32	40	29
With	70	82	92	93

The number of punished employees can be greater than one, if two employees choose the same minimum effort level, or it can even be zero, if all choose the same effort level, and thus, there is no *minimum effort*.[3]

The theoretical analysis of this team production game reveals that there is a multiplicity of equilibria if the game is played only once. In fact, each symmetric effort profile is an equilibrium: no employee has any incentives to change his effort level if everyone has chosen the same.[4] This mechanism of competitive exclusion generates a powerful reaction in the participants in an experiment composed of two blocks of ten rounds each, summarized in Table 6.2. Note that, in the last ten rounds, average effort levels reach 93% of the possible maximum.[5]

The competitive incentives system analyzed in the laboratory has double the advantages when compared to those based exclusively on punishments: it is more effective in raising effort levels, and they emerge as a consequence of the pressure the competitive mechanism exerts on employees.

However, the reader will notice that such a competitive system requires more information on individual effort levels than that based on random punishments. One must have at least ordinal information on efforts (to determine who is the employee exerting the least effort). In this sense this mechanism requires less surveillance than one punishing each employee for his (deviations from exerting) effort, as we would require cardinal information: that is, the absolute effort level for each employee.

The principal–agent model

If we try to analyze the problem of incentives from the firm's perspective – as an agent who also participates in the game – the horizontal scheme of team production is not useful. We must then go to the classic paradigm of studying work relations, which is the *principal–agent* model.

In its simplest terms, an individual (the principal) hires another individual (the agent) to perform an action or task for him. The key feature of this contractual relationship is that, after the contract is signed, an information asymmetry emerges. For instance, if the principal is a firm hiring a manager, it is possible that the firm's owner cannot observe the effort the manager exerts in performing his duties, as it is quite likely that the manager ends up having better information about the different opportunities the firm has.

These information problems must be foreseen by employers and especially by the principal, who faces the challenge of designing a contract that mitigates the adverse consequences of such information asymmetries.

The theoretical analysis of principal–agent problems, both when involving information asymmetries related to hidden information or to actions that cannot be observed (moral hazard), was the subject of an enormous research effort during the last quarter of the 20th century. As the fruit of such effort, a characterization of optimal contracts as functions of observable and verifiable variables was produced. Given that one of the observable variables for the firm is their production level, the optimal provision of incentives requires linking the employee's effort level to the acquired production level, establishing through this a piece rate payoff system. Hart and Hölmstrom (1987) is a classic reference, and Salanié (1997), Macho and Perea (2001) and Bolton and Dewatripont (2005) are recommended textbooks on the topic.

The fruits of experimental economics applied to labor economics had to wait until experimental methodology had become more definitely settled in economics. In fact, the first experiment on the basic principal–agent model, where the relationship between effort (real effort) and offered wage is examined, appeared in an accounting journal (Swenson, 1988). This experiment has a fixed salary for each unit produced and the treatment variable is the level of taxes. The main finding is that both labor supply and income (through taxes) are backwardly curved.

Up to the end of the nineties economists did not carry out experimental verifications in the laboratory. One of the most interesting findings is that monetary incentives – those provided by the experimenter – interfered with other types of incentives, natural to the situation that was being studied. Gneezy and Rustichini (2000) contrast the basic principal–agent model with real effort both in the laboratory and in field experiments. The basic idea of their paper is to verify a typical economic reasoning: if retribution depends on effort and people dislike effort but like money, then the higher the retribution, the higher the effort.

Gneezy and Rustichini do not observe such positive relationships between retribution levels and effort levels, but instead, when retribution is low, effort is lower than when nothing is paid (thus the title of the paper "Pay enough or don't pay at all"). More recent authors have confirmed the absence of a monotonic relationship between monetary compensation and effort (see Ariely, Gneezy, Loewenstein and Mazar, 2009).

However, perhaps the main contribution experimental economics has made to the study of the principal–agent problem is the overwhelming evidence favoring the existence of social preferences in the laboratory. As discussed in chapter 6 of Vol. 1, reciprocity and equality considerations influence experimental subjects' behavior. Before we analyze in detail the experiment that initiates this line of research, Fehr, Kirchsteiger and Riedl (1993) – FKR from now on – we analyze the work by Anderhub, Gätcher and Königstein (2002).

This work focuses on verifying the two main constraints a principal faces when designing a contract: *incentive compatibility constraint* and *participation constraint*.

a) The first one means that the principal must assume that the agent will exert the effort most convenient for him and will not take into account the principal's interests.

b) The second means that the principal extracts as much surplus as he can from the agent, up to the point where he is indifferent towards accepting and rejecting his contract.

In the experiment, the principal can offer the agent any linear contract, composed of a fixed wage and a variable payment, defined as a percentage of the firm's income, and at the same time he can suggest the agent's effort level. The extreme values are interesting:

• If there are negative fixed wages it means the agent pays the principal,

- While a 100% return rate means that all the firm's income is for the employee (this is equivalent to the agent being the owner of the firm).

Once the contract is offered, the agent can accept it or reject it (participation constraint) and, in case of it being accepted, the agent must decide on an effort level that determines the firm's income level but that implies a cost for the agent. The effort level suggested by the principal is simply that: a suggestion (incentive compatibility constraint).

In this experiment the efficiency contract has a very simple structure. As the only difference between the principal and the agent is that the latter decides the firm's income level (through his own effort), it is optimal to "make him the boss", that is, to "sell" the firm to him. Through this, the problem of incentives is resolved: the agent controls himself. The selling price is determined by the participation constraint.

The authors consider that, by effectively selling the firm, a 100% variable remuneration is observed in almost all cases, the effort suggested is efficient and the effort exerted by the employee is also efficient in almost all cases.

Up to this point everything is in order, but there are some important refinements. *i*) The first is that those agents receiving the most generous offers are the same choosing an effort level above the efficient level: that is, agents reciprocate the principals' good behavior. *ii*) The second is related to the fixed wage level. On this point, the principal does not extract the entire surplus from the agent, because frequently very high selling prices are rejected as they do not distribute the surplus from the contractual relation equally.[6]

The principal–agent behavioral model

According to Gary Charness and Peter Kuhn in their chapter in the "Handbook of Labor Economics" (2010), the experiment that has had most impact on labor economics is the *gift exchange* game by FKR. The theoretical idea originated ten years earlier, when Akerlof (1982) pointed out that work relations are incomplete (a contract is incomplete when it does not contemplate all the possible contingencies that may affect a work relationship). This typically occurs with the employee's effort level; if effort cannot be verified, incentives emerge for the employee not to exert effort in his job and, therefore, for the production volume to be inefficient.

Given that total inefficiency does not appear to be a characteristic that firms have in the real world, Akerlof argued that there must be voluntary cooperation between the firm and the employee: the firm pays the employee a wage above the necessary level and the employee returns the favor by exerting effort above the minimum level; this is, therefore, a favor exchange that yields a positive relation between efforts and wages.

FKR designed an experiment to verify the favor exchange theory designed by Akerlof. This is an experiment based on a two-stage game: in the first stage a three-minute oral auction is established, where firms bid for employees. Firms offer wages and any employee is free to choose the most convenient offer. If an employee accepts a wage offer, a binding contract between the employee and the firm is established in the second stage of the game. If a wage offer is not accepted, the firm is free, when there is time left, to increase their wage offer. After three minutes, the labor market closes and those economic agents who have not signed any binding agreement get zero. In the second stage, each employee has the task of *anonymously* exerting an effort level.

Regarding the experimental procedures, FKR randomly allocate the role of firm or employee among the experimental subjects. They run different sessions, all characterized by having an excess supply of employees. The identity of each couple, firm-employee, is anonymous, so that subjects do not know at any moment the identity of the person they are interacting with. In fact, employees and firms were seated in separate rooms, and communication between the rooms took place by phone. The main objective of this assumption is to avoid firms (employees) compensating for previous actions of an employee (firm). Finally, each pair of stages constitutes a period and the game is repeated for 12 periods to allow experimental subjects to learn.

Table 6.3   Scheme m(x)

X	0,1	0,2	0,3	0,4	0,5	0,6	0,7	0,8	0,9	1
m(x)	0	1	2	4	6	8	10	12	15	18

Let x_j be the effort exerted by employee j and let w_j be the wage per work unit accepted by employee j. The monetary costs associated to the employee's effort x are given by an increasing function $m = m(x)$ where $m(x_{min}) = 0$, as illustrated in Table 6.3.

There are two types of costs for the employee: the cost of going to work and the effort exerted. We denote the first cost as o and it is the opportunity cost of being at the work post; we denote the second as $m(x)$ and it is the cost of performance in the work post and, as we have illustrated in Table 6.3, it increases with effort. Therefore, the payoff an employee j with wage w_j and effort x_j receives is given by:

$$u_j = w_j - o - m\left(x_j\right).$$

For firm i, whose employee chooses effort x_i, its payoff is determined by

$$\pi_i = \left(v - w_i\right)x_i,$$

where vx_i is interpreted as the firm's income, in the sense that each unit of effort produces a unit of the good that is sold at price 1. In the experiment by FKR the values used were $v = 126$ and $o = 26$.

As is frequently the case in economic theory, we will assume that agents are maximizers and that rationality is common knowledge. We should think, therefore, in terms of the NE and, concretely in terms of the subgame perfect NE that arises in the environment used by FKR. Let us begin the second stage.

- Because effort is costly for the employees, and because the employees cannot be punished *ex-post* for exerting a low effort, there are no monetary incentives to choose an effort level above the minimum, x_{min}.
- Because the firm foresees the employee choosing a minimal effort level (regardless of the wage offered), it is rational for the firm to offer the opportunity cost of accepting the job, o.

Here it becomes interesting to understand why FKR add an oral auction between the firms to the basic principal–agent problem. The idea is to incorporate a job contracts' market and, given the excess supply of employees, to make this market competitive, and thus to converge to the competitive equilibrium ($w = 0$, $x = x_{min}$). As discussed in Chapter 1, the prediction and confirmation that a competitive experimental market converges to the competitive equilibrium is (almost) as old as experimental economics (Smith, 1964).

In the experiment by FKR no chance is given for gift exchanges to emerge: the pressure the competitive market has to set wages equal to the opportunity cost of working will make it unattainable for firms to provide favors to the employees who would later yield a return by exerting greater effort.

Table 6.4 presents the basic experimental results, regarding offered wages and offered efforts.

As can be observed, the experimental results show the existence of favor exchanges: there is an increasing relationship between wages and efforts, considerably above the competitive prediction (average wage was 72 and average effort was 0.4).

The main conclusion is that fairness and equity considerations survive in competitive markets, hindering wages from converging towards competitive levels. The firms expected that by offering higher wages employees would reciprocate with higher efforts. And this expectation was confirmed by the employee's behavior.

The presence of gift exchanges has been subsequently confirmed in other experiments, showing experimental robustness. Fehr, Kirchler, Weichbold and Gätcher (1998) observe this in the absence of competitive pressure – same number of firms and employees – Gächter and Falk (2002) observe it in repeated settings, where the same pair of firm–employee repeatedly interact in the gift exchange game. This is because, in the repeated game, selfish subjects imitate those who reciprocate. Maximiano, Sloof and Sonnemans (2007) observe it when there are multiple employees.

Table 6.4 The relationship between wages and effort

Salary	Average effort	Median effort
30–44	0,17	0,1
45–59	0,18	0,2
60–74	0,34	0,4
75–89	0,45	0,4
90–110	0,52	0,5

Conclusions

Experimental economics allows for the defining and solving of some of the basic questions affecting the establishment of wages and incentives by firms. We learn from the laboratory that, paraphrasing Nalbatian and Schotter (1997), a few drops of competition produce a very powerful effect on the behavior of subjects when participating in this type of study. While a vertical punishment system doubles team members' effort levels, a competitive system of exclusion, based on the threat of an identical punishment, causes effort levels to constantly approximate the optimum.

The existence of gift exchange within firms provides a complementary message. The increasing relationship between wage and effort suggests that firms can make use of the existence of social behavior patterns to design their remuneration systems. A relationship between employees and employers that accommodates offers based on explicit equity considerations, should ensure not only that it can survive in competitive environments, but also improve the firm's outcomes.

Notes

1. As the reader may have guessed already, it is a public goods problem.
2. This is the optimal situation for the team, for it allows the group profits to reach a maximum: the group profit if each member chooses an effort level $x_i = e$ is $N \cdot BNe$, greater than the one the group would get if no one exerts any effort ($N \cdot e$).
3. The logic behind this mechanism is that it only punishes those employees exerting the least effort for they exert less effort than others. If all make the same effort, no one makes less effort than anyone. Thus, even if all choose a very low effort level, the competitive mechanism of exclusion does not punish any team member.
4. The intuition is simple: no employee has incentives to change his effort level if all others have chosen the same effort level. If I increase my effort level in one unit, unilaterally, from any symmetric position, I individually assume the cost of my effort (1) and in return I receive a lower benefit ($B<1$); therefore I have no interest in increasing my effort. However, I am also not interested in decreasing my effort in one unit, because even though I increase my private benefits by one, I am punished for choosing the minimum level of effort, for which I lose all the group benefit.
5. Croson *et al.* (2015) analyze the effectiveness of excludability (exclusion of the lowest contributor) in a variety of team production teams. They find that excludability increases efforts, and is particularly effective in teams where the average or maximum effort determines team production.
6. Notice that this is the same conclusion observed in the ultimatum game (see chapters 6 and 8 of Vol. 1).

7
Experiments on Organizations
Jordi Brandts and Carles Solà Belda

Introduction

Organizations are crucial entities from an economic perspective. Multiple transactions take place in markets but many others occur inside organizations of all sorts or between organizations, where individuals may behave differently to how they would if interacting in markets. There is an extensive list of topics used to describe in detail organizational behavior: such as, the creation of organizations, organizational learning, innovation, organizational culture, communication and others (see for instance, Robbins, 2010). For this reason we will focus on two types of problems that organizations need to solve which have a fundamentally economic basis. These are problems of providing appropriate incentives to agents interested in the organization and problems of coordinating their actions. We will present, as much as possible, work relating to these matters along with other fundamental aspects of organizations. Taken together, these problems have the same objective: favoring the creation of value in the organization.

The provision of incentives is a classic problem that has been analyzed by researchers from different fields, with an accumulation of results that allows us to highlight two main aspects. The first relates to market differences found in research when compared to the standard model in economics, the principal–agent model. The second relates to the ongoing difficulty of integrating the social dimension of an organization which has incentive problems (Camerer and Malmendier, 2007).

Individual incentives and group work

Organizations are characterized by grouping resources and agents who (willingly) participate in activities that augment the generation of aggregate value and who share the wealth acquired from these activities. According to this account, from the perspective used by economics to study them since Alchian and Demsetz (1972), the group dimension is the key to understanding organizations.

In addition to acknowledging this group-related dimension of organizations, it is important to point out that organizations are structured on the basis of groups of people that:

• Perform activities with different objectives and,
• need to coordinate among themselves.

Team production exists because, at least in some situations, it is more productive to jointly use a group of workers as in an assembly line with specialization, or quality teams that enable knowledge transfer.

The interest in team production lies in its prevalence in firms and in the problems that arise in helping teams to function efficiently. The key theoretical reference is the work by Holmstrom (1982), further developed by others such as Holmstrom and Tirole (1989) in analyzing team production and illustrating the main problem: the impossibility of reaching an efficient production level as the outcome of a Nash equilibrium (see chapter 3 of Vol. 1 and onwards).

There are different solutions that can arise to classical problems in terms of more or less radical changes in the analysis framework. In this section we will briefly review some of these approaches (see also Salas, 1996) and the associated experimental findings.

First, we look at some classic proposals towards solving the efficiency problem in teams without changing the conditions of the model. The classic solution to the free-rider problems in team production (individuals who take advantage of others' work exert lower effort) is related to the supervision of a specialized agent who performs as the classic businessman (Alchian and Demsetz, 1972). Another solution to the group inefficiency problem relates to establishing rewards and setting certain objectives (or punishments). Here the prediction is a bit more problematic: the reward level must be such that the efforts required from the group members are maximal, otherwise there will be a multiplicity of equilibria, allowing some to stop working while forcing the rest to exert more effort (Holmstrom 1982). Another way to generate efficiency in team production consists of making workers post a bond. Examples of this type of mechanism can be seen in franchises.

Clearly, testing the finding by Holmstrom on the impossibility of achieving efficient team production is the aspect that stands out most in teamwork. Tests of these findings are linked to experiments on the prisoner's dilemma or the voluntary contribution game (VCG) given their strategic equivalence (see chapter 7 of Vol. 1 for a detailed analysis of these games).

Formally, team production can be presented as follows: Consider a group of N workers, where:

- each worker contributes, to the production, the effort e_i, $i = 1, .. N$;
- the opportunity cost for each worker is represented by $C(e_i)$;
- the resulting production function is represented by $Y = F(e_1, e_1, ..., e_N)$.

Each worker receives a payment according to the joint outcome, because the firm cannot observe their individual effort. In consequence, a standard way of establishing the compensation workers receive is $R_i(Y) = \left(\dfrac{1}{N}\right) * Y$, that is, the firm allocates to each worker a "symmetric" proportion of production.

The outcome mentioned above, where subjects behave as free-riders, is due to the employee's marginal income being lower than the marginal product, that is, what an employee receives marginally (for each additional unit of effort) is lower than his contribution to production.

The general experimental findings can be found in Davis and Holt (1993) and Ledyard (1995). For instance, the already classic works of Isaac and Walker (1988a, 1988b and 1991) and Nalbantian and Schotter (1997) obtain similar findings through different designs. Individuals tend to exert effort levels between 40% and 60% of their resources, although these outcomes change when interactions are repeated, reaching levels around 10%. Heterogeneity in opportunity costs does not generate different aggregate outcomes, according to Fisher *et al.* (1995).

In a different type of approach comparing the individual and collective performance of a task, Van Dijk *et al.* (2001) observe that a form of individual payment (a piece rate pay system) induces the same aggregate effort level to payment by teams. This finding is also observed in a different context in Vandergrift and Yavas (2011). In teams, the free-riding behavior is compensated by the excess effort exerted by some group members. Such over exertion ends up generating average effort levels similar to those generated by other systems of incentives.

There are different interpretations given to these findings. It appears that, in some way, predictions about inefficiency are observed only partially, although some of the proposed solutions have important effects (bonuses generate results), as appears to be suggested by the findings in Erev and Rapoport (1990) in other settings. One possibility is to think that the model does not adjust to the findings because it should incorporate some changes. The most common modification consists of thinking that real group interactions are repeated (and multiple experimental results have repetition, as we have discussed before). The traditional game theoretic results allow us to continue the theory-experiments discussion without much difficulty.

i. If we understand team technology as a game between the workers who participate, then when played only once there is a unique equilibrium (if we assume we are in a simultaneous or sequential game with complete information) that does not coincide with efficient production.

ii. If the game is repeated a finite number of times then the prediction based on the subgame perfect equilibrium (see chapter 3 of Vol. 1) allows us to state that the equilibrium of the initial game is repeated as many times as there are interactions. Thus, the theory predicts that finite repetitions do not change the prediction substantially.

iii. Kreps (1996) proposes a more imaginative alternative. This work shows that when an interaction is repeated a fixed number of times it is possible to achieve efficient production as the result of an equilibrium. The reasoning behind it consists of allowing the belief that one agent acts following a "cooperation and punishment" strategy. When facing this situation, it can be optimal for a rational player to play as a cooperator, exerting effort, thus maintaining the cooperative behavior of the "non-rational" agent. Such a strategy can sustain high levels of cooperation up to the last periods of the interaction, when the rational agent does not benefit from cooperating any longer as they are close to the end of the game.

This way of modeling the interaction may appear somewhat forced, but it is consistent with many findings when observing cooperation up to the last periods.

Another traditional way (this may be *iv*) of explaining cooperation both in team production games and in prisoner dilemmas is to represent the interaction as a game with infinitely many periods, each with a positive probability of the game ending (and players have positive discount rates). The findings show, once more, that it is possible in this type of interaction to sustain high levels of effort resulting from Nash equilibria (or subgame perfect equilibria). The argument here is that it is possible to build cooperative strategies that incorporate punishments for others' deviations. Such strategies may constitute equilibria that sustain cooperation, although there are multiple equilibria and, among them, non-cooperation can arise as well.

Another way of changing the environment is based on the thinking that, in reality, workers can acquire information about their partners' choices, even if this information cannot be verified. This could lead to considering teams as sequential games. In this case, there are also experimental findings in sequential prisoner dilemmas, as in Clark and Sefton (2001) or Solà (2002), and in sequential VCG that portray interesting dynamics, where reciprocity appears to have an important weight, causing high contributions.

We cannot conclude this section without addressing other models where individuals incorporate other motivations in their utility function. Cooperation in teams can be explained by "less strategic" arguments. As observed in chapter 6 of Vol. 1, social preferences (in many of their descriptions) explain the cooperative behavior observed in experiments with team technologies. As observed above, there is a wide range of models: inequality aversion by Fehr and Schmidt (1999) and Bolton and Ockenfels (2000), distributive preferences by Charness and Rabin (2002) or justice by Rabin (1993). We can add others, such as altruism by Rotemberg (1994) or reciprocity by

Bowles (1998). All these models include the notion that workers are not motivated exclusively by monetary gain.

Finally, there are additional factors that may have an important effect on cooperation between team members. Among these, organizational culture has been widely accepted and can be analyzed in a variety of ways: such as the pressure model by Kandel and Lazear (1992); or models incorporating identity notions, like that of Akerlof and Kranton (2005). Some of these proposals – to explain the preponderance teams have – find support both in the field (see Prendergast, 1999) as well as in the laboratory, where aspects such as communication and identity can be analyzed (Eckel and Grossman, 2005 or Chen and Li, 2009). Taken together, these proposals suggest that this route has the capacity to explain the preponderance of teams in organizations and the best way that they can be set up to ensure efficiency.

The coordination problem

The other fundamental problem of team production is the coordination between different members or components in an organization. As discussed in chapter 4 of Vol. 1, coordination is a central problem in game theory. In experimental economics coordination has been studied through order statistical games: the most popular of all the minimum games, first studied by Van Huyck *et al.* (1990). The minimum game or the "weakest link" game is used to represent an organization characterized by a production technology whose final production is determined by the person or unit that – due to a lack of skill or effort – contributes least to production. Kremer (1993, p. 551) describes the problem as follows:

> Many production processes consist of a series of tasks, mistakes in any of which can dramatically reduce the product's value. "Irregular" garments with slight imperfections sell at half price. Companies can fail due to bad marketing, even if the product design, manufacturing, and accounting are excellent.

This is an extreme case of production complementarities.

It has been suggested that the presence of complementarities could be the origin of multiple organizational problems. Some studies using data from specific firms analyze the effects such complementarities have. For instance, Knez and Simester (2002) study the successful change Continental Airlines in the mid-1990s. The critical element in Continental's success was the introduction of an incentives program designed to improve timelines, which at the end became a key determinant of the airlines' benefits. Knez and Simester (2002, p. 768) point to the importance of complementarities between different autonomous work groups to determine timelines: "When a flight departs late, employees, equipment, and terminal gates are unavailable to service other arrivals and departures. The problem is further compounded when flights carry connecting passengers because departing flights may have to be delayed to allow passengers to make their connections." Knez and Simester argue that the global nature of the incentives scheme by Continental played a central role in their success, guaranteeing employees that their greater efforts would coincide with the effort of their colleagues working in other units. In other words, a coordinated change was needed to improve Continental's situation.

Ichniowski, Shaw and Prennushi (1997) obtain similar findings in a study on the productivity of steel production plants. The type of steel production they study takes place in assembly lines where productivity is greatly conditioned to unscheduled downtime periods. This implies that an employee who does his job incorrectly (causing interruptions in his share of the assembly line) may well disrupt the efficiency of the entire chain.

It is very easy to characterize the minimum game. There are various players representing the components of the organization. Each player chooses simultaneously with the rest a strategy representing his effort level. The income each player receives is a decreasing function of his effort and increasing the minimum effort of some player or players. The following income function portrays the features just described, where *i* is any given component of the organization, *K* is a fixed income and the organization has *N* components:

i's income = $K - 5 * (i$'s effort$) + 6*($minimum effort of the *N* components$)$.

Notice that coordinating any common effort level is a Nash equilibrium, that is, a situation in which none of the components of the organization can unilaterally improve his income (see chapter 3 of Vol. 1). For instance, if the effort exerted by each component of the organization is ten, any of them would get a lower income if he was to increase or decrease his effort.

In particular, the situation where all the components of the organization exert the minimum possible effort is an equilibrium. Therefore, using the weakest link game as a metaphor it is possible to understand that some organizations or firms do not work well because they are stuck in a sort of trap where all the parts of the organization work considerably below their capabilities and none of them can improve the outcome on their own.

The main concern in this case is how to make a change. When a firm falls into a low performance trap, any attempt at improvement faces substantial obstacles – even if the benefits of a better coordination are evident – precisely because the low performance situation is an equilibrium: once they have fallen into the trap it is difficult to escape from it. The experimental findings of Van Huyck *et al.* (1990) show that many groups of people end up trapped in the minimum effort level and that, because of this, the problem is not only a potential one but real. Brandts and Cooper (2006) present a series of experiments designed to study to what extent changing a certain parameter in the incentives for the members of an organization may allow a group of people, trapped in a situation where all of them exert the minimum effort, to coordinate at a higher level of common effort, leading to an increase in everyone's income. In terms of the income function presented above, the question that arises is what would occur if the six, which multiplies the minimum effort by the N components, is replaced, for instance by a ten. This number may be interpreted as a fixed bonus paid by the managers of the organization.

It is easy to corroborate that, with this change, in theory the trap does not disappear, because if all the components of the organization continue exerting the minimum effort it would constitute an equilibrium. However, it is possible that this change motivates the components of the organization to attempt to escape from the trap, because a high common effort level is now more beneficial. Brandts and Cooper (2006) tested this by incrementing the bonus rates in different magnitudes. The findings notably enrich the problem:

i. A first finding of Brandts and Cooper (2006) effectively shows that increasing the bonus rate results in an increase in the minimum effort level.

ii. Second (and this may be surprising) there is no positive correlation between the magnitude of the increment of the bonus rate and its long-term impact on the minimum efforts. A higher bonus, level 14, does not generate higher minimum efforts. An interpretation of this finding is that what matters the most when an organization is badly coordinated is that the managers act by sending a signal showing that they aim to improve everyone's situation and not necessarily that the improvement in the size of the incentives needs to be very large.

iii. A third finding of Brandts and Cooper (2006) shows that the bonus rate can be reduced once coordination has been improved. This is important because, from the perspective of the managers in the organization, improving coordination requires paying higher bonuses. Therefore, it is important for the organization to know that the bonus may be reduced without this returning the minimum effort to its original level.

Brandts and Cooper (2006) study the role communication plays between managers and components of the organization in solving coordination problems. Work in the area of organizational behavior indicates that communication is one of the crucial variables that influence change (see, for instance, Ford and Ford, 1995 and Kotter, 1996). In fact, there are reasons to believe that communication will be particularly effective in organizations affected by coordination failures, as it helps in positively influencing the beliefs held in the different units of the organization. The capacity to do so may be seen as an essential characteristic of leadership.

The experiments by Brandts and Cooper (2007) study the effects of changing the available channels of communication between managers and employees. In the baseline treatment, managers only control financial incentives and communication is not possible. In two other treatments communication is allowed:

- unidirectional – managers can send messages to the employees; and,
- bidirectional – managers can send messages to the employees and vice-versa.

The content of communication between managers and employees is completely free, that is, participants may send any message they want subject only to some minimal constraints. One of the main aims of this work is to systematically analyze the impact that the content of different types of messages has. That is unusual in economics and links this work with organizational studies and with psychology. All messages were recorded and their content was quantified using a systematic coding structure, a common methodology used in psychological studies with verbal protocols as well as in previous economic experiments including communication (for a similar analysis see the last section in chapter 5 of Vol. 1).

The aim of this study is not only to establish that communication is a valuable tool for managers, but also to explain how communication improves profits. The questions asked are the following. Do more channels of communication lead to a higher minimum effort level, keeping the financial incentives fixed? Which communication strategies are more effective in increasing the minimum effort? What is more important to increase profits: the choice of financial incentive or the choice of communication strategy?

The main conclusion of this study is that communication between managers and employees may play a key role in escaping coordination failures. More concretely, the effective use of communication helps experimental firms to increase the minimum effort, where bidirectional communication between managers and employees is greater than unidirectional communication from the managers to the employees. More effective communication is more valuable when increasing the managers' benefits than a manipulation of the employees' bonus rate.

Not all messages between managers and employees have the same beneficial effect. The most effective strategy the managers have appears to be very simple and, in the end, natural. Managers should request a concrete effort level and underline the mutual beneficial effects of a high level of effort. The aim is to act as a good coordination mechanism. It is useful to highlight to what extent employees are well compensated, in spite "of the unrelative unimportance" that they are especially well compensated. For the employees the most effective message is to advise the manager, thus giving the firm the benefits of having more than one person thinking about sorting out collective problems.

Leadership

Leadership may be one of the ways in which the two main problems of organizations can be solved, the provision of incentives (in the wide sense) and coordination. In this section we will point to some of the aspects that have been experimentally studied in relation to forms of leadership and the effects they have.

Leadership is a widely researched topic in the field of the organization of firms, distinguishing, first of all, between the characteristics leaders have and their behavior. This vision was subsequently abandoned to focus on concrete aspects of the relationship between leaders and their collaborators (reference groups) and the optimal forms of leadership. Also, some specific types of leadership have been defined: such as transitional leadership or charismatic leadership (this idea is very influential in managerial environments). The experimental literature on leadership has concentrated its efforts on concrete aspects of behaviors that result in effective leadership, or its possible impact on the results of the organizations.

Leadership is a key aspect of the functioning of organizations. Particularly when an organization is experiencing difficulties, it is crucial to have a leadership that encourages the organization to overcome the crisis. There are already a number of works analyzing this question: all such studies use a simple game to represent the organization and analyze different aspects of leadership.

A series of works study the role of leadership in the context of public goods games or prisoner dilemmas. In these types of context leaders face the challenge of getting others to do something they would not do in the leader's absence. The key to leadership is getting the other members of the organization to put aside their immediate interests to promote the wider interests of the group. One mechanism by which leaders can influence others is to lead by example. Concretely:

- In sequential prisoners dilemma experiments it has been observed that those who decide on a second place frequently cooperate if the person deciding first cooperates, but they hardly ever do it if the first mover does not cooperate (Clark and Sefton, 2001).
- Gächter *et al.* (2012) in sequential public goods experiments corroborate that the contributions of the followers increase with the contributions of the leaders. In this case, the behavior of the

followers can be seen simply as another expression of the cooperation condition that has been observed in multiple contexts and which can be described through the now standard models of Fehr and Schmidt (1999) and Bolton and Ockenfels (2000).

- Brandts *et al.* (2015) also study sequential public good games with leadership and analyze different communication strategies that may help to improve the contributions of the followers, exceeding what is known as the restart effect. Communication is by far the more effective method, when compared to advice by externals on the benefits of reciprocating the leader's contribution.

Given the behavior of the leader, it is possible to argue that the followers mechanically follow his motivations. On the contrary, the leader must make his choice before knowing if the others will follow him and, therefore, it can be assumed that by being in such a position he needs an additional emotional push that consists of getting conditional cooperation to start moving. Note that in this type of situation a leader who decides to cooperate could be motivated both by social elements and by strategic considerations. Gächter *et al.* (in press) study in detail whether leaders who are reciprocal cooperators are better leaders than the individualistic ones. The results of the experiments show that people who in another task had been cooperators, contribute more to the public good than the individualistic ones, and end up being better leaders.

Arbak and Villeval (2007) go one step further and study the endogenous emergence of leadership as well as its consequences. The context is – as in the previous study – a sequential two-person public goods game where the participants can choose whether they want to be first or second in making choices. The results show that, although it is on average costly, a significant proportion of participants volunteer to act as the leader, making the first choice of the game. The choice of leading is mainly influenced by the sex of the person, as well as by other personality traits, such as generosity and open-mindedness. Another finding is that the motivations leaders have are diverse. Among the cooperator leaders there are people with altruistic motivations: some wish to teach others to get a better outcome for the group. There is also evidence suggesting that some people are sufficiently concerned with maintaining a good public image to want to be considered effective leaders.

The work of Brandts, Cooper and Fatás (2007) studies leadership in coordination problems that, once again, is located in the context of conditional coordination by the weakest link. In a situation in which the components of an organization differ in terms of their capacities, the question asked is: who will take the initiative to get the organization out of the difficult situation it is in? The findings are surprising and show that the most capable individuals are not the ones assuming leadership, but those with the most usual characteristics, that is, those who belong to the most numerous group. This effect may be due to some sort of group identity or to the cognitive simplicity of working together with similar people.

Conclusions

The conclusions we wish to draw from the presentation of this series of experimental studies is that the experimental method can be successfully applied to the analysis of some key problems in the functioning of organizations.

This methodology has also allowed the testing of hypotheses that were previously functionally impossible to analyze, given the lack of internal information from organizations. In addition to the topics of incentives, coordination and leadership, mentioned above, other important topics that have started to be addressed with experimental method are those of: creative processes and organizational growth, organizational learning and processes of organizational change, among others.

8
Macroeconomic Experiments
Francisco Lagos and Ernesto Reuben

Introduction[1]

Unlike microeconomic models and game theory, which frequently aim to obtain generalizable results, macroeconomic models are usually built to analyze very concrete aspects of reality and are seldom generalizable to other fields.

For many years the controlled manipulation of macroeconomic variables to understand the effects of institutions or alternative policies was considered, in practical terms, to be impossible. Therefore, many considered that macroeconomic matters could not be addressed with laboratory experiments. However, laboratory methods are today increasingly used to answer macroeconomic issues. This change has been due partly to changes in macroeconomic modeling and partly to improvements in the technology used to design more complex laboratory experiments.

Below we summarize some of the most relevant laboratory experiments in the field of macroeconomics, across three different areas. The first part focuses on monetary economics. The second explores some relevant aspects of international trade. Finally, the third provides a discussion on the use of laboratory experiments to test macroeconomic policies.

Monetary economics

Experimental studies on monetary economics are based on the different uses of money in market economies. It is argued that money plays three roles: first, as a store of value; second, as a medium of exchange; and third, as a unit of account. In this section we summarize multiple experimental studies designed to investigate theories related to each of these different uses of money.

Money as a store of value

Experiments exploring the role of money as a store of value aim to understand questions such as: how can an object with no intrinsic value be used as a store of value? And, how can the optimal price and quantity of such an object be established, given that it does not have a consumption value on its own?

A good experimental study exploring these questions is Hens *et al.* (2007).[2] The authors focus on testing whether an object that plays the role of money can achieve a stable value. Their study is closely related to a well-known case where 150 couples exchanged baby-sitting duties with one another (for a detailed description see Sweeney and Sweeney, 1977). The benefits from their agreement are obvious:

- The couples who are not planning to go out on a certain night can easily baby-sit for another couple's children;
- This allows other couples to have a very well-deserved night out.

Clearly, for the agreement to work, there must be a system protecting it from any abuse. For this reason, the organizers introduced a natural solution: they issued coupons equivalent to an hour of baby-sitting. If couple *A* baby-sits for couple *B*, then *B* pays *A* in coupons which, afterwards, *A* may use another day to get any available couple to baby-sit for them. In other words, they create their own currency.

When the system was launched, the organizers surprisingly found that it was prone to collapse. On the one hand, if they issued too few coupons, couples would tend to hoard them (that is, they would save too much), which as a result led to a low demand for baby-sitting and a collapse of the system (a recession). On the other hand, if they issued too many coupons it resulted in excess demand for baby-sitting and a dramatic decrease in the amount of baby-sitting hours couples were willing to offer for a coupon (inflation).[3]

In the experiment of Hens *et al.* (2007), in each period, subjects' preferences for a perishable good are randomly determined (either with a strong or a weak preference) and they must decide whether they want to buy or sell units of the good. To buy units of the good, a subject must have coupons (this constraint is precisely what gives coupons value), and sales of goods increase a subject's holdings of coupons. The unique prediction with rational expectations and an infinite horizon is that a subject opts to buy goods depending on his preferences in the current period as follows:

- Case 1: If a subject has a strong preference, he always buys units of the goods.
- Case 2: If a subject has a weak preference, but his coupon holdings are sufficiently high, the subject also buys units of the goods.
- Case 3: If a subject has a weak preference and his coupon holdings are *not* sufficiently high, the subject sells units of the goods to acquire more coupons.

It is straightforward to show that there is a unique optimal quantity of coupons that maximizes the number of trades possible and, therefore, social welfare. The authors used the amount of coupons in the experimental economy as the manipulation variable. In general, Hens *et al.* (2007) reported that the theory is widely corroborated: subjects' strategies coincided well with the strategies described above. Furthermore, exogenous increases in the total amount of coupons in the economy led at first to an increase in the volume of trade. But eventually, as it continued increasing, it was followed by a stark decrease in the demand for coupons, because subjects were not interested in accumulating any more coupons. Finally, the amount of coupons from which the volume of trade starts decreasing corresponds with the optimal quantity of coupons predicted by the theoretical model. This experiment nicely illustrates the difficulty central banks face in determining an optimal quantity of money in the economy.

Money as a medium of exchange

As a medium of exchange, money must serve as a store of value, but clearly there are many other objects that are stores of value but are not media of exchange. Therefore to understand the role of money, it is especially important to understand why other objects with higher rates of return do not substitute for money as a medium of exchange.

The overlapping-generations model is a well-known environment that provides money with a role both as a store of value and as a medium of exchange (Samuelson, 1958). Camera *et al.* (2003) use this model to investigate whether money is substituted as the medium of exchange when there is another object, a bond, that can play the role of storing value and that also bears interest (the bond was conceived so that it paid certain dividends in each period and had no terminal date). The equilibrium prediction in this context is that individuals will exclusively use the object offering the highest rate of return (the bond) as a medium of exchange and will abstain from using the other object (money). However, Camera *et al.* (2003) propose two complementary hypotheses to explain why some individuals could continue using money in this context:

- Accumulation;
- Habit.

The first is the accumulation hypothesis, which establishes that bonds are hoarded and not used as media of exchange because people want to receive the bonds' dividends. This hypothesis is tested by comparing two treatments; one where bonds are traded before dividends are paid (i.e., the subject buying the bonds gets the dividend) and another where bonds are traded after dividends are paid (i.e., the subject selling the bond gets the dividend). If the accumulation hypothesis is true, there must be more subjects using money as a medium of exchange in the treatment where bonds are traded before dividends are paid.

The second is the habit hypothesis, which establishes that subjects use money instead of bonds because "old habits die hard." This hypothesis is tested by comparing two treatments: one where subjects first play with money as the sole store of value before bonds are introduced, and another where both money and bonds are introduced from the beginning.

Camera *et al.* (2003) find substantial support for the habit hypothesis: money coexists with bonds as a medium of exchange in treatments where subjects begin with money as the sole medium of exchange and bonds are introduced afterwards. In addition, in line with the accumulation hypothesis, it is more frequent for money and bonds to coexist when dividends are paid after bonds are traded. If dividends are paid before bonds are traded, and both money and bonds are introduced simultaneously, subjects exclusively use bonds as the sole medium of exchange.

Money as unit of account

Money's role as a unit of account is uncontroversial. Clearly, prices are typically quoted in terms of money units and not in terms of, say, olives. However, this poses a problem, as money typically depreciates in value over time due to inflation, while, generally, the value of products, such as olives, is kept constant. To avoid this problem, most macroeconomic models presume that, in their transactions, economic agents evaluate all choice variables in real terms: that is, they are not subject to money illusion. However, data from surveys (Shafir *et al.*, 1997) or simple introspection suggests that this assumption does not always hold. Experimental studies of money as a unit of account do not only study whether some individuals are prone to money illusion, but also how well they assess the consequences money illusion has on the behavior of prices in markets.

Imagine a consumer who finds, to his surprise, that his salary has doubled overnight, but he lives in a country where, like his salary, all prices have also doubled. Will the consumer feel richer today and behave differently than yesterday? The traditional assumption suggests that, because the salary increase is purely nominal and, in real terms, there is no change, the consumer will not change his behavior. However, experimental studies by Fehr and Tyran show that thinking in nominal terms is common and that, in some circumstances, it can have noticeable effects on market prices.

Consider Fehr and Tyran (2001) as an example.[4] In this experiment subjects repeatedly interact in a game where they compete in an oligopolistic market (see Chapter 2). In each period, a subject's income depends on his chosen price and the average price chosen by the other subjects. The market was designed so that it has a unique equilibrium and, importantly, there are strategic complementarities in the choices subjects make. In other words, the optimal strategy for each subject has a positive relationship with the average price chosen by the other subjects, so that if the average price increases, subjects have incentives to increase their own price.[5] Because the market demand function is mathematically complex, each subject simply received a table indicating his income for each price he may choose and for each realized average price. In this way, it was not complicated for subjects to find their optimal strategy.

The main purpose of this experiment is to see how subjects react in this market to a nominal *shock*.

1. From period 1 to 20, subjects first play with a table where the equilibrium price is 18 points.
2. From period 21 to 40, all subjects get a new table where the equilibrium price is six points (the nominal *shock*).

Even though prices have changed in nominal terms, because the incomes are relative to the average price, in real terms subjects are in the same situation. Fehr and Tyran (2001) design four treatments. In the first treatment, subjects receive tables containing prices in real terms, so that it is a trivial task for them to calculate the optimal strategies before and after the nominal *shock*. In the second, subjects receive tables containing prices only in nominal terms; thus, they must exert a little more effort if they want to calculate the optimal strategy after the nominal *shock*. Treatment differences in subjects' behavior after the nominal *shock* can be attributed to money illusion.

Finally, the third and fourth treatments are identical to the first and second, except that subjects play against computer "players," knowing they have been programed so that they always play optimally in real terms.[6] By using virtual players with a pre-programed strategy, the authors make sure that the subjects know that other players do not suffer from money illusion. By doing so, it is possible to disentangle the effect when subjects suffer from money illusion from the effect when subjects believe that others (but not they) suffer from money illusion.

The experimental findings show that, in three of the four treatments, after the (fully anticipated) nominal *shock*, prices are immediately adjusted to the new equilibrium. Only in the treatment with nominal income tables and human players did this not occur. In this treatment, price adjustment is considerably more sluggish (see Figure 8.1).

These findings are interesting because they suggest that even when subjects have no problems converting their nominal incomes into real incomes (when they play against computerized players there is no difference between the real table and the nominal table treatments), money illusion can have prominent effects on prices, in markets with strategic complementarities, simply because subjects believe there are

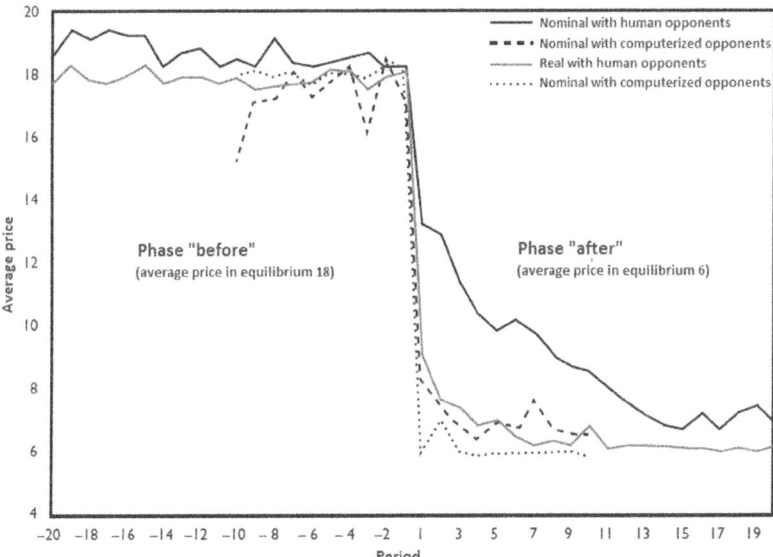

Figure 8.1   Money illusion in Fehr and Tyran (2008)

other subjects playing as if nominal incomes were the same as real incomes.[7]

In summary, money illusion has important effects on market prices when participants have incentives to "follow the crowd." Nonetheless, while money illusion and the market's strategic environment are interesting explanations for nominal price stickiness, most macroeconomists point to other sources to explain this phenomenon, including friction in the acquisition and dissemination of information or costly price adjustment. Experimental studies that investigate the relevance of these other mechanisms will be an important source of future research.

Our knowledge of the way money acts in the economy is fundamental, but in spite of this, our knowledge of how the uses of money are affected by multiple variables is still limited. Some of the questions for which we do not have fully satisfactory answers include: what is the optimal amount of money in the economy and when and why is there hyperinflation? What are the effects when money is seen in nominal terms by some but not by all economic agents? The experiments discussed in this chapter provide plausible answers to some of these questions, but there is still a lot of work to be done.

International trade

Another field of the macro-economy in which experimental methods have played a relevant role is that of international trade. Noussair, Plott and Riezman (1995) (see Chapter 1) conducted the first experimental attempt to create and study some of the most relevant features of international trade.

In an exchange environment guided by multiple interacting markets, the main objective of this ambitious study was to experimentally draw a distinction between aspects such as the comparative advantage law, factor price equalization, production efficiency and the effect of taxes on international transactions. The authors consider two environments:

- The first one, motivated by the *Ricardian* model of international trade: labor is the only input.
- The second is one where both capital and labor are used as inputs in production.

In both environments there are two countries and within each country two types of agents: consumers and producers. There are equal numbers of consumers and producers in each country (four consumers and four producers).

In the first environment, consumers own the only production factor, L, and have induced preferences to consume the final goods labeled as Y and Z. Producers have, as well, an initial endowment of L to produce and sell the final goods Y and Z. Additionally, all agents may attempt to obtain benefit by speculating both with their inputs and their outputs (final goods). Production factors are not mobile across countries and the final goods Y and Z can be exchanged between them. The two countries differ only in their production technologies. Country 1 has a comparative advantage in the production of good Y, and country 2 has a comparative advantage in the production of good Z. Table 8.1 contains the main experimental parameters for both environments.

Table 8.1 Experiment Noussair, Plott and Riezman (1995)

		Scenario 1	Scenario 2
Provisions	Consumers, country 1	L1 = 2	L1 = 5, K1 = 3
	Consumers, country 2	L2 = 2	L2 = 3, K2 = 5
	Producers, country 1	L1 = 1	L1 = 0, K1 = 0
	Producers, country 2	L2 = 2	L2 = 0, K2 = 0
Production technology	Country 1	$Y = 3L, Z = L$	$Y = L, Z = K$
	Country 2	$Y = L, Z = 2L$	$Y = L, Z = K$

The economy works as follows:

- Consumers sell their initial endowment of L to producers in their same country, and afterwards they buy units of Y and Z produced by either country.
- Consumers gain utility (money) both by consuming as well as through the benefits generated from speculation.
- Producers in each country buy L from the consumers in their country, and they can use L to produce Y and Z, which subsequently they can sell to consumers from either country.
- The produces obtain utility (money) both from their production activities as well as from the benefits generated from speculation.

While some experimental sessions allowed for free international trade others imposed a tax for trading between countries.

Capital, K, is added to the second environment as an input to production and both countries have identical linear production technologies but different endowments of capital and labor. In environments 1 and 2 there were six and eight markets operating simultaneously, respectively. Each variable had its own market (for instance, the final good $Y1$ produced in country 1 had its own market). These markets were implemented using computerized double auctions (see Chapters 1, 3 and 4).

The main hypothesis tested in environment 1 is the law of comparative advantage. The competitive model predicts, for this law, that countries 1 and 2 will exclusively produce goods Y and Z, respectively, and each country completely specialize in exporting the good it produces. In accordance with the competitive model, the prices of the two goods should be equalized across countries and the prices of inputs should equalize to their marginal productivities. Such predictions can be contrasted to the inefficient outcome under autarky, where there is no trade between countries and, hence, specialization does not emerge.

In environment 2, the competitive model predicts that both countries will produce both final goods. However, in accordance with the model, country 1 will specialize in exporting good Y, and country 2 will specialize in exporting good Z. Under free-trade conditions, prices of the goods are equalized across countries, implying convergence in the prices of the factors. Such equalization does not take place in the autarkic model.

The main finding from this pioneering work is to experimentally observe, for the first time, that the law of comparative advantage accurately predicts patterns of commerce and trade. In the *Ricardian* environment there is nearly complete specialization by producers in the two countries, and in the environment with capital, the two countries are net exporters of the good for which they have a comparative advantage. In general, the qualitative predictions of the model are confirmed. Convergence processes are observed and such convergence occurs faster for quantities than for prices. According to these experiments, there is not much support for the autarkic model.

Using a simpler design, Noussair, Plott and Riezman (1997) conducted the first experiments exploring the behavior of the economy in international finance markets. More concretely, the aim of these experiments was to better understand the ability of the competitive equilibrium model to predict and control prices and exchange rates.

Once again, there are two countries and each of them produces two final goods Y and Z.

However, unlike the previous experiment, there are no longer any factor inputs or production processes. Each country was populated by six subjects:

- Three of whom were sellers of (endowed with) good Y and buyers of good Z.
- The other three were sellers of (endowed with) good Z and buyers of good Y.

In addition, each buyer is indifferent, in terms of utility, between acquiring Y or Z in his home country or in the foreign country. All subjects were endowed with a large amount of cash (only) in their home currency, and foreign country purchases required acquisition (cash) in advance of the foreign currency. Therefore, in this economy there are two countries with six agents each, two goods, and two types of currency, which are only valuable to agents in their own country. Subjects' preferences were induced to value both goods and the home currency only (the end-of-session redemption value of any foreign currency holdings was zero). In each country, markets in the two goods and foreign currency were implemented using computerized double auctions (see Chapters 1, 3 and 4).

Further restrictions designed to force the use of the international finance market were imposed. First, buying and selling in a country must take place using the local currency. Second, no agent was allowed to export but all agents were allowed to import. That is, in order to buy in the other country, agents must use the financial market and purchase foreign currency using their local currency. Once they had purchased goods in the other country, they could transport them to their country without cost and, once there, they could either consume them or re-sell them in the local currency.

The exchange rate – that is, the price of the currency from country 1 in terms of the currency from country 2 – was determined so that the balance of payments would be in equilibrium. Thus, the exchange rate could equate the demand and supply for currencies in both countries, arising out of the flow of international transactions. Given this, the main hypothesis tested concerns the law of one price (which guarantees purchasing power parity). The alternative hypothesis is again, as in the autarkic model, where the no-trade outcome is realized, and the law of one price does not hold.

The authors conclude that these experiments solidly reject the autarkic model. They find that, in general, most aspects of the competitive model work fairly well, but some do not. For instance, they find that exchange rates quickly converge to the equilibrium values predicted by the competitive model, although the prices for some goods do not converge. On the other hand, the law of one price fails to obtain. The authors' conjecture is that this failure does not arise because of the competitive equilibrium model, but because of different speeds of convergence of prices in the two domestic markets.

Fisher (2001), in a subsequent paper, revisits the issue of the law of one price and purchasing power parity by constructing a greatly simplified version of the experiment in Noussair, Plott and Riezman (1997). In Fisher's design, there are:

- Two types of goods: green and red.
- Two types of currencies: green and red.

There are two countries and each produces only a single good. The green goods are available at an elastic supply at a constant price (in green currency) in each period, and the red goods are available at an elastic supply at a price (in red currency) announced at the beginning of each period. In essence, the supply of goods in the market is perfectly controlled by the experimenter. In addition, red and green currencies are exchangeable in the market.

Even though each subject was endowed with a large supply of the green currency, in each period there was a fixed (and, therefore, perfectly inelastic) supply of red currency in the market. After the price of the red good was announced, the supply of red currency was auctioned off in a "second-price, sealed bid auction" (see Chapter 3). The market-clearing price (equal to the second lowest bid submitted) of a unit of red currency in terms of green currency was interpreted as the nominal exchange rate for that period.

Once the exchange rate was determined, subjects were free to buy units of green and red goods. Fisher's main hypothesis was – a relative version of purchasing power parity – that the nominal exchange rate is constant between periods. A second hypothesis was – absolute purchasing power parity – that the real exchange rate equals the marginal rate of substitution between foreign and domestic goods.

Fisher finds convincing empirical evidence for both the relative and absolute versions of purchasing power parity. This finding confirms the conjecture of Noussair, Plott and Riezman (1997) as to why they do not find support in their experiments for purchasing power parity. That is, the divergence in convergence of prices of goods in both domestic markets appears to be relevant and must be considered when designing laboratory experiments.

Macroeconomic policies

Because it is normally impossible (and ethically questionable) to experiment with real macroeconomic policies, the laboratory provides an ideal environment to examine the possible impact macroeconomic policies may have before they are applied. In this section we discuss two areas where laboratory experiments have been used.

Credible commitments

An important practical macroeconomic policy issue concerns the way to overcome problems related to the use of discretionary policies, which are optimal (for the policy makers) in the short term, but not in the long term. A clear example of this problem arises in models where policymakers have incentives to create inflation aiming to reduce unemployment (the well-known Phillips curve). Kydland and Prescott (1977) show how these types of discretionary policies generate a situation where the policymakers ratify the inflation expectations of the citizens, resulting in an excessive level of inflation and no improvement in unemployment.

If the policymakers were able to credibly commit to a zero-inflation policy, the problem would be avoided and the social optimum could be implemented. In theory, Barro and Gordon (1983) solve this problem by modeling the situation as an infinitely repeated game between the policymakers and the citizens. In their model, players use strategies so that the policymakers have a reputation that allows them to implement the socially optimal policy (as in all infinitely repeated games, many other equilibria also exist). The experiments of Van Huyck *et al.* (1995, 2001) were designed to test these theoretical ideas.

Van Huyck *et al.* (1995, 2001) use a game that captures, in a very simple way, the three situations mentioned above. They are concerned with policymaking in situations where the policymakers:

- Have no way to commit.
- Are able to credibly commit.
- Are able to credibly commit to maintain their reputation in an infinitely repeated game.

In each stage of the game there are two periods and two players. In each repetition, subjects are randomly assigned roles as either policymakers or citizens.

In the first period of a repetition, the citizen is endowed with an income, Y, and must decide how much of this to consume in this period, $C_1 \geq 0$, or invest, $I \geq 0$, at a return rate, $r > 0$. The amount available for consumption in the second period, $C_2 \geq 0$, depends on the investment in period 1 and the fraction transferred to the policy-maker through a tax rate, m, concretely, $C_1 = (1 - m)(1 + r) \times I$.

In the treatment simulating the situation without credible commit-ments, the policymaker chooses the tax rate *after* the citizen has made an investment choice. In this case, the optimal choice for the policy-maker is to choose the highest tax, $m = 1$, and therefore it is optimal for the citizen to invest nothing, $I = 0$.

In the treatment simulating the situation with credible commit-ments, the policymaker chooses the tax rate *before* the citizen has made an investment choice. In this case, the policymaker has the incentive of choosing a lower tax rate to stimulate the citizen to make a positive investment (in fact, the policymaker chooses the tax rate maximizing the social welfare: $m^* = r / (1 + r)$ and $I^* = Y$).[8]

Finally, in the treatment simulating the infinitely repeated game, subjects play indefinitely without any credible commitments: they repeatedly interact in fixed pairings and have a sufficiently high prob-ability of continuation so that there is an equilibrium supporting the social optimum.

The findings verify higher investment levels and, in general, they are closer to the social optimum in treatments with credible commit-ment than in those without them (or without repetition), which are closer to the no-investment equilibrium. Treatments with indefinite repetition show intermediate investment levels.

In other words, the authors find that reputation is an imperfect substitute for a credible commitment mechanism. This is an important finding because outside the laboratory we do not, generally, have good mechanisms to make policymakers commit and, instead, we settle for reputation-based mechanisms.

Fiscal policies

Riedl and van Winden (2001, 2007) design an experiment to examine if unemployment benefits can generate vicious cycles of unemploy-ment and cause deterioration in the general economy.[9] Concretely, they experimentally study how the economy works in countries where unemployment benefits are financed by a tax rate applied to labor income, as in many developed countries. They consider two types of economy: closed and open. In the open economy there are two countries, one is relatively small and the other is relatively large.

There are consumers and producers in both economies. Each consumer is endowed with K units of capital and L units of labor, which they can sell to the producers as inputs of production. In addition, for each unit of unsold labor consumers get an unemployment benefit. They can use the unemployment benefits to buy in the final goods market. Consumers obtain utility (money) from the two final goods, Y and Z and from "leisure": that is, from the unsold units of labor.[10] The goods Y and Z are produced in two separate sectors. Producers in these sectors need K and L as inputs, which they transform into final goods through their production technologies.[11] The technologies for the two goods differ because the production of X is relatively more dependent on capital, while the production of Y is relatively more dependent on labor. Producers gain utility (money) from what they sell (once the production cost has been discounted). The cost of labor includes a tax rate proportional to the wage.

- There are four markets in the closed economy: two factor markets (for K and L) and two final goods markets (for Y and Z).
- The same markets are present in the open economy, but both the one for capital and the one for the final good Y are open markets (exposed), while the one for labor and the one for the final good Z are domestic markets (protected).

In both economies, markets were implemented using computerized double auctions (see Chapters 1, 3 and 4). In addition, while the number of consumers and producers was the same for both countries in the open economy, the consumers in the *large* country were endowed with seven times more units of K and L than consumers in the *small* country.

The authors find experimental evidence supporting the negative economic effects of using taxes applied to wages as a means to finance unemployment benefits. In addition, they find that employment can be promoted using the budget deficit. However, once the wage tax is forced to adjust for the budget to be in equilibrium, both the level of real GDP and other economic indicators, tend to slowly stabilize to a substantially low level that does not reach the equilibrium prediction.

Van der Heijden *et al.* (1998) test a possible explanation for the stability of the social security system through a voluntary "social contract" between successive generations. The authors design an individual decision-making mechanism over transfers in an experimental environment between overlapping generations, in which the current generation can monitor and react to transfers made by the previous generation. With this aim, in one treatment they provide subjects with information about the level of transferences (pensions) of the previous generation; and in a second treatment this information is not given. In both treatments, each subject (generation) P_t decides how much to transfer (pension) to subject P_{t-1} and, likewise subject P_{t+1} decides how much to transfer to subject P_t, and so on. Subjects live for two periods. In the first period (when they are young), subjects are endowed with nine units, from which only seven can be transferred. In this period, the *young* subjects decide how much to transfer to the current *old* subjects. The units the young subjects do not transfer are used for their own consumption. In the second period (when they are old), subjects receive a non-transferable endowment of one unit plus the units transferred to them by the current *young* subjects. In addition, subjects had induced preferences for a stable consumption in both periods.

The main finding in these experiments suggests that the level and stability of the transfers system does not rely on the possibly of controlling (monitoring) transferences from previous generations. That is, the availability subjects have to maintain transfers from the young to the old seems to be independent of the possibility they have to know about the choices previously made. In addition the authors find scant evidence supporting the effect of rewards or punishments between generations.

Conclusions

Empirical studies of macroeconomic models are notably hard to conduct. In general, researchers cannot directly observe the behavior of economic agents and they can only infer, indirectly, the effects of macroeconomic policies. In addition, evidence that is consistent with a particular theory can also be consistent with other alternative theories based on very different assumptions. Laboratory experiments provide the advantage of convincingly discarding alternative theories, without forgetting they have the necessary limitation of studying very simple economies (in comparison to a real economy). By way of contrast, empirical data is much richer but leaves the debate open as to the relevance of different theories. Given that both means of research have advantages and limitations, it is best to use them as complementary tools. For instance, experimental findings can be used to reinforce the interpretation given to empirical data and conversely, empirical research can inspire new laboratory experiments.

Notes

1. We want to thank John Duffy for his excellent review of the literature on macroeconomic laboratory experiments, contained in a chapter that will appear in the next volume of the *Handbook of Experimental Economics, Volume 2*, edited by John H. Kagel and Alvin E. Roth, under the title "Macroeconomics: A Survey of Laboratory Research." This work has been a guide for the elaboration of our chapter.
2. McCabe (1989) and Deck *et al.* (2006) use similar designs to study the role of money as a store of value.
3. Due to the multiple problems mentioned above, the number of families partaking in the agreement has decreased from over 250 in the 1970s to less than 20 in 2010.
4. Other articles in this area include: Fehr and Tyran (2007, 2008) and Noussair *et al.* (2007).
5. A good example of a market with strategic complementarities is an oligopolistic market, in which agents compete in quantities.
6. Additionally, instead of playing for 40 periods, in these treatments subjects only played for 20 periods (ten periods before and ten after the nominal *shock*).
7. In a subsequent article, Fehr and Tyran (2008) show that this effect is due to strategic complementarities. That is, they demonstrate that even with human players, if there is strategic substitution in the market, the adjustment of prices is very quick.
8. Letters marked with a star m^* and I^* refer to the values of m and I that maximize the players' incomes keeping in mind that both of them act optimally. In other words, in the equilibrium of the game.
9. This research project was developed for the Dutch Ministry of Social Affairs and Employment. The authorities specifically requested the authors to develop laboratory experiments to assess the formulation of their macroeconomic policies.
10. The consumer's preferences are induced by a linear-logarithmic version of a Cobb-Douglas utility function.
11. The production functions are discrete approximations of a CSE (constant substitution elasticity) production function.

9
Experiments in Political Economy

Humberto Llavador and Robert Oxoby

Introduction

Voting and elections play dual roles as social choice systems. On the one hand, they act as a preference aggregation system: they are used to choose between different alternatives when citizens do not agree on their preferred choices. On the other hand, they act as an information aggregation system: when individuals share the same preferences but each has only partial information on the *state of the world*, a voting system can be used to aggregate the decentralized information, increasing the probability of choosing the best alternative.

Experimental studies have analyzed both aspects of elections and have explored the same issues as empirical researchers in political science: electoral participation, voters' strategic behavior, convergence of electoral platforms, retrospective voting,[1] coordination between voters when there are more than two alternatives, the importance of information transmission, etc.

In this chapter we address some of these topics through representative experiments in political economy, the branch of political science with a formal theoretical framework most similar to economics. Clearly, we cannot cover all the existing literature. In particular, we omit experiments on "committee decisions," which in many cases lie behind the experiments we present here.

The first two sections study the role of elections as preference aggregation mechanisms, analyzing voters' and candidates' behavior. The third section studies the capacity of voting to aggregate information, presenting some of its most important implications and paradoxes.

Voters' behavior

Abstention and participation

A prominent and well-studied topic on voters' behavior is known as the *participation paradox*. In elections with a large number of voters the probability that a single vote will be decisive or pivotal is close to zero. Hence, from a cost–benefit perspective, it is irrational to vote. That is, since one single vote is almost surely irrelevant to the outcome of the election, any small voting cost (for instance, the trade-off of not going to the beach) is greater than the expected benefit.[2]

Therefore, from a rational choice perspective, the task is not to explain the relatively "low" levels of participation in large elections but, on the contrary, to explain why so many citizens decide to participate in such elections when each individual's vote has, most likely, no effect on the electoral outcome. Do voters ignore the strategic calculation of voting, which is based on the probability of being the decisive voter? Do they vote simply because they like to?

Empirical analysis of these questions using field data is difficult because the variables of interest, in particular perceptions of benefits and participation costs, are almost impossible to measure or approximate using observable variables. Thus, laboratory experiments present a particularly appealing option. But, how can we design the costs and benefits of participation in elections?

In their experiments, Bornstein (1992) and Schramm and Sonnemans (1996a, b) identify the act of voting with the procurement of tokens for a participant's group. Participants are divided into two groups competing against each other to choose the winning option.

- Each individual decides how many tokens to purchase, at a given cost, knowing that her payoff will depend on the total number of tokens acquired by her group compared to the number of tokens acquired by the other group.
- All members in a group get the same payoff.

With this design it is possible to explicitly manipulate the cost of voting (through the price of each token) and the benefits of voting (through the payoffs contingent on the number of tokens acquired by the group).

Despite the simplicity in their design, the experiments by Schramm and Sonnemans allow for the analysis of the effects, not only of the costs and benefits of voters' behavior (that is, the number of tokens each individual purchases), but also of variations in group size.

Furthermore, since the authors were able to directly characterize voters' choices, they were able to manipulate the institutional framework to explore the effects of different political institutions on voters' behavior. With regard to institutions, Schramm and Sonnemans considered two systems:

- a majority system (where there is only one winning group); and
- a proportional representation system.[3]

In the majority system only the members of the winning group receive a positive payoff, while in the proportional system each individual receives a payoff proportional to the fraction of total tokens acquired by her group. Each voter participated in 20 rounds, and the experiments were conducted with groups of 12, 24 and 48 voters in each election.

The results of the experiments are surprising. First, while participation behavior responds as expected in terms of costs and benefits (participation increases when the cost decreases and decreases when the cost increases), participation was consistently higher in the majority system than in the proportional representation system.[4] The authors also found evidence suggesting that participation (i.e., the number of tokens acquired) decreases with group size, although the effect was small and statistically insignificant.

In these experiments, costs were kept constant for all individuals, suggesting that changes in behavior for different institutional frameworks and group sizes are fundamentally due to participants' perceptions regarding the benefits of voting. In other words, these findings suggest that voters are more sensitive to perceived benefits than to voting costs.[5]

An important problem in many voting models is the existence of multiple equilibria and the high degree of strategic uncertainty among participants, complicating the interpretation of the findings (see chapter 3 of Vol. 1). Levine and Palfrey (2007) avoid this problem by using a different design. In their experiments, all members of one party (or group) obtain the same benefits, but have different costs. Each individual's cost is private information, but individuals know the distribution of costs in their group. This design allows the experimenter to choose the distribution of costs to yield a unique equilibrium.[6]

The theoretical model behind this experiment is presented in Palfrey and Rosenthal (1985). Although we do not provide a description of the model in this chapter, we present three of its hypotheses that are tested against the experimental findings.

i. The *size hypothesis*: keeping preferences constant, participation should decrease as the number of voters increases.
ii. The *competition hypothesis*: keeping the number of voters constant, participation should increase as the fraction of voters supporting each party comes closer to 50%.
iii. The *underdog hypothesis*: supporters of the minority party (the party supported by less than half of the population) should vote in a greater proportion than those in the majority party.

The findings in the experiments of Levine and Palfrey (2007) provide support for all three hypotheses, although participation tends to be greater than theoretically predicted in elections with a larger number of voters, and lower than theoretically predicted in elections with a smaller number of voters. The findings are presented in Table 9.1, comparing the participation rates predicted by the theoretical equilibrium with those observed in the experiments.

Table 9.1 Participation rate: balance and data

Size matches (majority party–minority party)	2–1	5–4	6–3	14–13	18–9	26–25	34–17
Minority participation in equilibrium	54	41	46	27	30	21	24
Real participation of the minority (data)	53	44	48	38	38	33	39
Majority participation in equilibrium	64	37	45	23	30	17	23
Real participation of the majority (data)	64	40	45	28	36	27	36

Note: Reproduction of Table in Palfrey (2009). Source: Levine and Palfrey (2007).

Strategic behavior

Besides the analysis of participation, experiments also provide an opportunity to investigate to what extent voters behave strategically and choose to support candidates who, despite not being preferred, can help obtain a more favorable outcome in the elections. Strategic voting can have important consequences. For instance, consider a majority election with three candidates, A, B, and C, with the following characteristics:

- Most voters (say 60%) prefer A or B to C, but they are equally divided between the two candidates: 30% support A and 30% support B.
- The remaining 40% prefer C.

Without strategic voting, if all individuals were to vote for their favorite candidate, C would win the election, even with the majority preferring any of the other two candidates.[7] However, if those voters supporting B voted for A, A would win the election and those voters preferring B would be better off than if C were elected. Notice that strategic voting is not enough to avoid the situation of choosing the least preferred candidate since voters are required to coordinate on which candidate, among their preferred ones, they will support.

Forsythe *et al.*(1993) presents a procedure that is common in this type of experiment: with three political alternatives, participants (voters) are assigned to "voting groups" which share the same preferred option. In this way, the experimenter can manipulate payoffs, group sizes and, in particular, the degree to which the majority is split between the two alternatives with the largest support.

Overall, experimental results show that a Condorcet loser may be elected in the absence of signals indicating how individuals will vote. In these circumstances, voters, ignoring the distribution of preferences, may vote for their favorite candidate or not vote at all.

However, when there are informative signals about the distribution of preferences (e.g., through surveys or election campaigns supporting one of the majority candidates), a majority of voters behave strategically and vote for the candidate with the largest support, even if she is not their favorite candidate.

Elections and political competition

The median voter and two-candidate elections

The median voter theorem (MVT) establishes that in two-candidate majority elections, under certain conditions electoral platforms will converge to the preferred option of the median voter. In other words, if there are Condorcet winners, they will be elected. But does this actually happen? At first sight electoral platforms appear different for different candidates or parties, even in majority voting systems with a single winner. That is, not even under the appropriate theoretical conditions (a majority winner system and a two candidate election) do the political platforms converge to the preferences of the median voter.

In order to analyze this question in the laboratory, McKelvey and Ordeshook (1982) designed a series of experiments with two candidates in a two-dimensional policy space.[8] Each experiment consisted of a series of rounds where two subjects, always the same or "partners" (see chapter 7 of Vol. 1), chose strategic paths after observing the choice of the other candidate in previous rounds and the consequent electoral outcomes.

Thus, candidates were the only participants in these experiments: they had no preferences over the political issue and received benefits only if they won the election. Voters were actors simply voting for the candidate whose proposal was closer to their ideal policy. Finally, the configuration of preferences and the distribution of voters were such that there was a unique Condorcet winner.

Figures 9.1a and 9.1b summarize the policy proposals of the candidates for the first five and last five rounds in a five-voter election. One can observe that, in this two-dimensional policy space with a unique winner, candidates converge to the point of the Condorcet winner. A comparison of the graphs shows the process of learning and adjustment over time. Similar to the convergence of prices in a competitive market, electoral platforms in competitive elections converge to the Condorcet winner.

These experiments were conducted in a setting of complete information. Candidates (the experimental subjects) were fully informed about the preferences of all voters and it was assumed that voters (artificially created) voted based on the strategies (i.e., proposals) of the candidates.

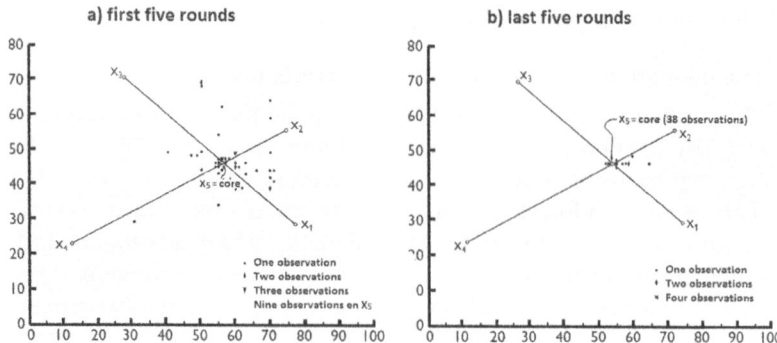

Figure 9.1   Spatial elections with five voters and equilibrium existence.
Reproduction of Figures 5.8a and 5.8b in McKelvey and Ordeshook (1990).
Source: McKelvey and Ordeshook (1982).

However, in reality candidates do not have complete information but only incomplete data on the preferences of voters. In addition, voters usually receive relatively poor information on the proposed policies that candidates will implement. Therefore, it is very relevant to know how much information on voters' preferences and candidates' policies is necessary to obtain convergence to the Condorcet winner.

Regarding information about voters' preferences, Plott (1991) conducted ten experiments where candidates did not know voters' preferences but could ask voters questions and periodic surveys on voting intentions were conducted.[9] The main conclusion in these experiments was the replication of previous findings, supporting the idea that the information provided by surveys is sufficient for the convergence of candidates' proposals. However, pilot experiments "without questions" suggest that convergence would not have happened in the absence of surveys.

Regarding the poor quality of information that voters have during elections, McKelvey and Ordeshook (1990) conducted a series of experiments in a one-dimensional policy space. Each experiment consisted of eight elections with candidates perfectly informed about voter preferences. Each session started with candidates choosing policy positions, followed by two opinion surveys, and finally, voting to determine the winner. Subjects were either informed or uninformed voters.

- A majority of voters (half plus one) were *informed* voters, who received information about the position of the candidates.
- The rest were *uninformed* voters who had to base their decision on signals like endorsements, opinion polls or their relative position compared to other voters.

The design was such that the information received by the electorate as a whole was enough to produce a rational expectations equilibrium where all voters voted as if they had perfect information on the candidates' positions.

In the experiments, uninformed voters voted "correctly" 84.9% of the time and candidates converged quickly to the preferred position of the median voter. This is evidence of the robustness of the median voter theorem in environments with incomplete information.

In other words, in a one-dimensional political system with office-motivated candidates (i.e., those deriving benefits solely from being in office) we find support for the hypothesis of total revelation of voters' information and the convergence of candidates' platforms.[10]

Although subjects participated in a series of rounds to learn the information processes, the results so far have restricted the outcomes to a single election. Alternatively, a large literature studying voting behavior suggests that voters may vote retrospectively – i.e., based on the previous policy choices of a politician in office. A large set of experiments address this issue by linking the payoff of voters with the policy position chosen by an incumbent candidate.[11]

In the absence of electoral campaigns, voters have to choose between re-electing the incumbent and choosing an alternative candidate, based exclusively on the flow of benefits received in the past. The incumbent only knows the history of policy positions chosen by previous candidates in power and the voting support they received. He must choose his policy position with this information. Findings from these experiments show once more that, on average, candidates converge to the position of the median voter, in spite of the limited information.

In summary, even for *laboratory elections* where the relevance of voting outcomes is limited, when candidates are identical and office motivated the median voter theorem is an acceptable approximation for the outcome of a two-party, spatial competition election: candidates' strategies tend to converge and the Condorcet winner is a good predictor of the outcome of competitive elections.

A second relevant implication is that the convergence result for two-candidate elections is robust to the information that voters have about candidates and the information that candidates have about voters. In fact, very little information is necessary. Therefore, it may seem irrational for voters to accumulate costly information when there are surveys, endorsements, word-of-mouth transmission, or other sources of information that are essentially free.[12]

However, as we will see in the section on *asymmetric competition*, such convergence outcomes are not immune to changes in the framework and disappear in the presence of differentiating characteristics independent of the candidates' policy positions (*valence issues*), or when candidates have policy preferences and there is uncertainty regarding the position of the median voter.

Elections with more than two candidates

Many elections are characterized by more than two candidates competing under a simple majority system. This sort of election is especially interesting because there is no conclusive theoretical result. In fact, practically anything can happen in equilibrium. Even in a three-candidate election with a Condorcet winner, a candidate choosing the Condorcet position may be crushed on both sides by the competition of the other two candidates and end up losing the election. Therefore, it is possible that Condorcet winning positions are not chosen.

Plott (1991) uses the experimental framework described in the previous section to compare elections with two and three candidates. He finds that experiments with three candidates tend to reproduce equilibrium outcomes observed with two candidates. Candidates tend to converge, although the variance is greater in elections with three candidates. However, the dynamics that result in this equilibrium are hard to predict, as it is difficult to separate the effects of candidates from those of voters, and there is evidence that voters are sophisticated and behave strategically (see the previous section on *strategic behavior*). It is possible that outcomes are sensitive to the moment when the experiment ends, that is, to the length of the election. Competition leads candidates to the median voter's position, but the candidate in the middle, squeezed by the other two candidates, changes his policy, triggering a new cycle and a new convergence process. In summary, sophisticated voting and the time when elections end seem to explain the final outcomes.[13]

So far it is the experimenter who has assigned the role of candidate and, therefore, the number and the characteristics of candidates were given exogenously. *Citizen-candidate* models solve this limitation and are a good framework for analyzing the nomination process, as well as endogeneity in the number of candidates (see Osborne and Slivinski, 1996, or Besley and Coate, 1997). Consider a set of *n* citizens (the experimental subjects):

- Each citizen is assigned an ideal point within an interval.
- Subjects decide whether to run as a candidate or not.
- Running as a candidate implies a cost and candidates can only run with their ideal point as their policy position.
- Payoffs are the result of the decision to run as a candidate or not, the net benefit of winning the elections (that is, the benefits of wining the elections minus the cost of being a candidate), and the distance between the ideal policy point and the winning policy.
- Voting is an automatic action, and the winner is the candidate whose ideal point is closest to a majority of voters' ideal points. In case of a tie the winner is chosen using a lottery. If no one chooses to be a candidate, a point is randomly chosen among all the ideal points.[14]

The results are revealing. When the cost of running as a candidate was high (so that the net benefit of winning was low), 86% of the observed decisions were consistent with the theoretical prediction that only one participant would present himself as a candidate, which also coincided with the position of the median point.[15]

When the cost is low, the theory is inconclusive and predicts multiple equilibria. The parameterization used in this experiment could generate equilibria with one, two or three candidates. However, the experimental results supported, in general, a symmetric equilibrium with two candidates and differentiated positions, converging slowly to equilibrium: it is only during the last five rounds that we observe a larger number of candidates from citizens on both sides of the median and a smaller number of candidates in the median position.

Asymmetric competition

Elections feature candidates with asymmetric characteristics, that is, candidates differing in aspects other than their policy positions. Common sources of asymmetry include an incumbent position, personal valence issues or ideology. The most relevant implication of these asymmetries is that the theoretical equilibrium now predicts divergence in the policy positions of the candidates, even when the policy space is one-dimensional.

For the sake of presentation, we will focus on the case where candidates differ in an identifiable personal characteristic (i.e., a valence) but results are similar when candidates hold a political ideology or have preferences over the different policy positions.

A candidate's personal valence, such as her last name, being a movie or sports star, or her physical or personality traits, is relevant because it offers her an advantage or a disadvantage over the other candidate (independent of the support received from her spatial policy location). Consider the following experiment:

- Two candidates simultaneously choose a position between three possible alternatives: $\{L,C,R\}$.
- Candidates ignore the exact location of the median voter, but they know that the median voter is in C with probability a and that it is in L or R with the same probability, i.e., $(1\text{-}a)/2$.

Probability a is the control variable in the analysis. The experimental design is such that personal valence makes candidate 1 the advantaged candidate and hence she wins the election when both candidates are located at the same distance from the median voter.

In this context, Aragonés and Palfrey (2004, 2005) find support for the following theoretical implications: the disadvantaged candidate tends to locate in more extreme positions than the advantaged candidate; moreover, as the distribution of voters polarizes (the value of a decreases), the disadvantaged candidate shifts towards the center, while the advantaged candidates shifts towards more extreme positions.[16]

These outcomes suggest that, in the presence of personal valence and uncertainty on the location of the median voter, candidates not only diverge in their policy positions but, in general, candidates with a relative disadvantage will tend to diverge when the distribution of voters' preferences is unimodal.

Information aggregation and voting

In addition to choosing between different options or candidates, voting can serve to aggregate information distributed across different individuals. In the standard model of information aggregation, a group of individuals share common interests and have to choose an alternative that grants different payoffs depending on the *state of the world*. The group does not know the state of the world, but bits of information are distributed across individuals.

The question is: can a voting system aggregate such decentralized information and find the true value of the state of the world, allowing the group to make the best choice?

Condorcet's jury theorem

The original information aggregation model can be found in Condorcet (1785), and the virtues of majority voting are summarized in the well-known Condorcet's jury theorem (CJT). In a few words, this theorem states that, under very general conditions, majority voting efficiently aggregates the information individuals have and is able to discover the truth, or to choose the adequate option, as long as individuals vote sincerely. For instance, if each individual has the correct information with a 60% probability, the perception of the majority will be correct more often than the perception of any particular individual. In particular, three individuals with statistically independent information who decide through majority voting will choose correctly 64% of the times, against the 60% when choosing individually. Furthermore, this probability tends to one as the group size tends to infinity. The 60% probability is just an example, but the results hold for any probability larger than 50%, that is, whenever individuals hold valuable information. Therefore, CJT asserts that majority voting is a good information aggregation system.

Experimental analysis has demonstrated the robustness of CJT, as the experiments of Ladha *et al.* (2003) and of Guarnaschelli *et al.* (2000) show.[17] Ladha *et al.* (2003) conducted the first experiments studying whether majority rule leads to an improvement against individual decisions. In those experiments a group comprised of three individuals had to discern the color of a ball drawn out of an urn filled with 60 white balls and 40 black balls. Without further information, the best option for each individual is to state that the ball is white and to be correct 60% of the times. In order to analyze information aggregation, the experimenter offered individuals a signal.

Before stating a color, each subject would draw a ball form a second urn, whose composition depended on the color of the initial ball:

- if the initial ball was black, the second urn would only contain black balls;
- if the initial ball was white, the second urn contained 60 white balls and 40 black balls.

Therefore, if an individual drew a white ball from the second urn he would know that the color of the initial ball was white. On the other hand, if he drew a black ball, the initial color could have been either white or black.[18] After receiving the signals, the three individuals in a group would decide the color by majority voting, that is, the color with at least two votes would be chosen.

Ladha's experimental results confirmed the potential for majority decision to aggregate information. Groups deciding by majority voting chose the correct color of the ball 93.75% of the times. Surprisingly, this result is not only higher than the theoretical rate of correct individual choices (76%), but also higher than the rate predicted by CJT (78.9%).[19]

The swing voter's curse

When abstention is an option, the presence of uninformed voters can result in what Feddersen and Pesendorfer (1996) coined "the swing voter's curse:" an uninformed voter may decide to abstain (either by not participating or by making his abstention public) with the intention of delegating his decision to those voters who are "better informed."

This is referred to as a curse because a vote is relevant only when it is decisive, and uninformed voters are decisive when voting against informed voters. If an uninformed voter is decisive he may hinder the choice of the informed voters sharing his same interests. It is rational, therefore, to abstain, even when voting is not costly. On the other hand, if there are *partisan* voters (voters who always support an alternative, regardless of the situation), it may be rational for an uninformed voter to vote, and do so even against his a priori information, to counter the votes of partisan voters, leaving the choice in the hands of *independent* and informed voters. This rational behavior requires a sophisticated reasoning. Are voters sophisticated enough?

Battaglini *et al.* (2010) explore this problem in the laboratory. In their experiments, two urns contain 75% of white balls each, but they differ in the color of the remaining 25% of balls. They are red in urn *A* and yellow in urn *B*. The experimenter randomly chooses one of the urns. Individuals know the probability of choosing each urn, and must guess which urn was selected. Before deciding, each individual draws one ball from the chosen urn, observes its color and puts it back in the urn. Hence,

- some citizens become informed, when they draw a red or yellow ball; while
- others remain uninformed, when they draw a white ball.

In the experiments, uninformed voters abstain 91% of the time, supporting the swing voter's curse hypothesis. Moreover, in the presence of *partisan* voters who always vote for *A*, uninformed voters show rates of participation and unconditional voting for *B* in line with the theoretical predictions. That is, uninformed voters rarely vote for *A*, while the frequency of votes for *B* increases with the number of *partisan* voters.

Conclusions

In this chapter we have presented a few representative examples of experiments in political economy. We have not been exhaustive, neither in the list of experiments nor in the topics in political economy studied by experimental economics. Many other important topics, such as the role of communication media, expenditure on electoral campaigns or the possibility of government coalitions, have also been studied in the laboratory.

Nonetheless, despite the small number of experiments covered in this chapter, we can draw two conclusions. First, there is clear evidence supporting the existence of voters' strategic and sophisticated behavior, observed for example in the electoral participation choice or in the swing voter's curse experiment. Second, although with some quantitative differences between outcomes in the laboratory and theoretical predictions (e.g., electoral participation is usually larger in the laboratory), experimental results tend to corroborate most predictions from the theoretical models, such as the potential for majority elections to aggregate information or the divergence of policies when candidates have differing ideologies or personal characteristics.

Notes

1. Citizens vote retrospectively when they evaluate politicians in office through their past decisions.
2. The participation paradox was already identified by pioneering authors in the formal study of political theory, such as Downs (1957) or Riker and Ordeshook (1968).
3. The proportional system is dominant in Western Europe. In this system various members of Congress are elected for each electoral district, with each party receiving a number of representatives approximately proportional to the number of votes received. On the other hand, Anglo-Saxon countries, such as the United Kingdom, New Zealand, Canada and Australia, tend to use a majority system, in which only the representative with the highest number of votes in each district gets elected, leaving the rest of parties without representation.
4. Participation was around 50% during the first rounds of the majority system, decreasing consistently to 20% in the last rounds. On the other hand, initial participation in the proportional system was around 30%, gradually decreasing to 20% in the final two rounds. Nevertheless, it is important to stress that it is not possible to know if participation would have continued decreasing with experience if the experiments had continued for higher number of rounds.
5. After the 20 official rounds, Schramm and Sonnemans allowed participants to communicate with others in their group, being able to exchange their impressions about the outcomes in previous rounds. Afterwards they conducted a series of "surprise rounds" voting. In these rounds participation practically doubled, suggesting that communication (and involvement in general) may play an important role in determining electoral participation.
6. The equilibrium is characterized by critical values for costs so that:
   – all those members with costs above the critical value for their group abstain; and
   – the rest of the group, whose costs are below the critical value, vote.
7. A *Condorcet loser* is a candidate who always loses in an election against any of other candidate. In our example, candidate C is a Condorcet loser since in any pairwise election 60% of the electorate would vote against him. Similarly, a candidate is a *Condorcet winner* if he never loses in a pairwise majority vote.
8. When a vector of characteristics represents policies, we say that the policy space is multidimensional (either because the policies are intrinsically multidimensional, for instance a progressive taxing system, or because the

vector refers to a package of policies, with each dimension representing a different issue). When there is a single issue and this issue is characterized by a single parameter (for instance, a proportional tax rate), we say that the policy space is one-dimensional.

9. Candidates usually asked questions of the sort: "How many of you would prefer that I propose point *x*?"

10. However, when the analysis is extended to two dimensions, the convergence of candidates is slower. One may argue that multidimensional policy spaces imply substantially more difficult decisions and that candidates would end up converging if the number of rounds were increased. Nonetheless, such conjecture still requires to be proven.

11. See, for instance, Collier *et al.* (1987).

12. Findings in Collier *et al.* (1987) and Williams (1991) support this view when they extend the experimental environment by offering voters the possibility of acquiring, at a cost, information on the candidates' positions.

13. Indeed, only in one out of the 11 experiments is it the candidate in the middle who wins the election.

14. The details of the experiment, which was conducted with five individuals in 10 sessions, can be found in Cadigan (2005).

15. In this experiment, as it is frequently observed in *entry games* (games where there is a binary choice: to enter or not to enter), there is an excessive number of entries compared to the theoretical prediction of the Nash equilibrium.

16. These findings are robust, as they have been replicated for different subject pools and various instruction protocols.

17. Both articles find similar conclusions, although the main focus in Guarnaschelli *et al.* (2000) is on unanimity decisions, which we do not cover here.

18. In fact, if the drew a black ball, the initial ball was black with probability 62.5%. This number follows from Bayes theorem (see chapter 2 Vol. 1):

$$\Pr\left[Black \mid \text{signal\_Black}\right]$$

$$= \frac{\Pr\left[\text{signal\_Black} \mid Black\right] \times \Pr\left[Black\right]}{\Pr\left[\text{signal\_Black} \mid Black\right] \times \Pr\left[Black\right] + \Pr\left[\text{signal\_Black} \mid White\right] \times \Pr\left[White\right]}$$

$$= \frac{1 \times 0.4}{1 \times 0.4 + 0.4 \times 0.6} = 0.625.$$

19. These values obtain from calculating the probability of success conditioned to the signal. An individual following the information from his signal would succeed with the following probability:

(Pr[{white,white,white}|*White*]+3x[{white,white,black}|*White*])xPr[*White*]+ (Pr[{black, black, black } | *Black*] + 3 x [{black, black, white } | *Black*]) x Pr [*Black*]= 0,648 x 0,6 + 1 x 0,4 = 0,789.

10
Field Experiments and Development Economics

Francisco Alpízar and Juan Camilo Cárdenas

Introduction

In this chapter we discuss field experiments in the context of development economics. Given that field experiments are addressed in this chapter for the first time in the book, we start by defining them. Experimental economics was born fundamentally out of laboratory experiments, so it is natural to build our definition based on the existing difference between field and laboratory experiments.

Note that differences between field and laboratory experiments are essentially methodological and not conceptual, and as such the conceptual framework developed in the previous chapters of this book is applicable to both types of experiment.

Our second aim is to analyze the methodological and logistical particularities of field experiments in economics, particularly when applied to topics and locations relevant to a developing country. Our third aim is to briefly review the main topics covered up to the present, proposing new topics, which can still be explored. Field experiments can be a powerful tool for studying the economic and political processes developing countries go through, such as weak governments with little capacity to implement policies, incomplete decentralization of public policies and decisions, extreme inequality in income and opportunities, and high dependence on natural resources, among others.

Finally, our fourth aim is to analyze the influence that field experiments have had and could have on the field of public policy at all levels, from the central government through to local communities. Although most of our experience and examples take place in Latin America, we have aimed to write a text with wider geographical applicability.[1]

Field experiments: definition[2]

The transition from experiments in the laboratory to conducting them in the field is motivated by the researcher´s interest in gaining greater contextual relevance, one of many important features of good experimental design. Unavoidably, researchers have found it important to make the design more realistic, because testing the design in the field is in itself interesting, or because they seek to explore to what extent the findings of laboratory experiments can be generalized when in less controlled situations.

However, the growing interest in field experiments should not be interpreted as the exhaustion of the laboratory approach. On the contrary, the methodological difficulties as well as the loss of control inherent to any field experiment, its higher costs and its limited replicability – for instance, because there is less access to participants sharing similar characteristics – make them a poor instrument for an initial exploration of hypotheses.

We suggest researchers start in the laboratory, exploring multiple treatments that over time will allow us to refine the field experiment design. Moreover, this may even allow us to link some potential results in the field to the typical context of the laboratory, thus facilitating the generalization of findings and inferences. In addition, the sometimes complex findings from a field experiment may require examination in a context with greater control, which would imply returning to the laboratory. Thus, laboratory and field experiments are complementary.

In their influential article, Harrison and List (2004) describe six factors that determine whether an experiment is a field or a laboratory experiment:

i. The type of *participants*: in the laboratory, participants are usually university students. Participants in the field come from more heterogeneous socio-economic groups, or from particular "target" groups (artisanal fishermen, for instance).
ii. The type of *information* framing the decisions: unlike the laboratory, where decisions are often made in an abstract framework,

a field experiment can recreate a reality that, in its constituent elements, is not too different from the reality participants experience in their daily life. For instance, the experiment may use information on the production function of a particular group, or the ecological features of a fishing system, to create a realistic environment that allows and promotes participants to use their "life" experience as a guide to make their decisions in the game.

iii. The type of good or *service* exchanged or "produced" in the game: instead of being abstract, participants can focus on a realistic good or even on one that is specific to their own experience (for instance, the number of hooks per fishing trip).

iv. The types of *decisions, rules or institutions* framing the experiment: it is common that rules and institutions in an experiment are designed to mimic existing rules in the population being studied, and that decisions made by participants are realistic enough for them to think of their behavior in a real context.

v. The means of *payment*: although participants are typically paid both in the laboratory and the field, payoffs in the latter tend to be closer in form and amount to the reality experienced by the participants and they can even be non-monetary (for instance, fishing equipment or home-related goods, equivalent to the average daily income, to compensate at least the opportunity cost of participating).

vi. A wider *context* of the experiment: because the design of a field experiment represents an effort to come closer to reality, it can be expected that participants bring with them their usual behavioral patterns in society, which can range from male chauvinism to peer solidarity.

For instance, as part of an interaction process between communities, between 2001 and 2002 academics, NGOs and local authorities conducted a set of experiments linked to the problem of resource exploitation in the mangrove forest on the Pacific coast of Nariño in Colombia (Candelo *et al.*, 2002). The findings of the experiment were presented to the community to contribute to the discussion of the challenges and possibilities arising from the self-management of natural resources. The games conducted helped in creating a space for discussion to build voluntary agreements regarding the responsible exploitation of the mollusk known as Piangua (*Anadara tuberculosa*).

Typical patterns of behavior in a society are an obvious determinant of any economic agent's decisions and, in and of themselves, can be an object of study; for instance, when comparing the cooperative behavior of fishermen belonging to different communities.

The six factors mentioned above can be combined in multiple ways within a field experiment. In their paper, Harrison and List propose a typology of four types of experiments:

i. Conventional *laboratory* experiments, using students in an abstract context; not linked to from the particular reality of the participants.

ii. *Artefactual* field experiments, which are the same as conventional laboratory experiments but with a nonstandard subject pool.

iii. *Framed* field experiments, either in the characteristics shaping the decision or the institutional and informational framework available to the participants.

iv. *Natural* field experiments, where the environment in which participants make their choices is not different to their natural environment and where the subjects do not know that they are in an experiment.

There are two types of experiments, with increasing recent relevance, left out of this taxonomy. On the one hand, *randomised control trials* (Banerjee and Duflo, 2009), which are basically oriented towards the construction and implementation of public and private policy programs, originated from a random design which separates a control group from other groups assigned to different treatments.

On the other hand, there are *natural* experiments, which occur unplanned and do not require previous designs. Unlike field experiments, the difference between natural experiments and laboratory experiments is not only methodological, but also conceptual, moving away from the topics previously discussed in this book. For this reason, we will not address either of these type of experiments in this chapter.

Methodological features of experiments applied to development economics

Types of participant

The main reason for doing experiments with subjects that differ from the typical student pool is to come closer to the population of interest. In many cases financial reasons or the very objectives of the research require a target population, so that the decision to work with a non-standard subject pool precedes the experimental design.

In the context of most developing countries, working directly with a sample of the population of interest is particularly important because the population of university students tends to be systematically different. As an illustration, note that the group of people who gain access to university education in Latin America is small, 23% of the young people between 18 and 24 years old in 2001 (World Bank, 2002), and they are fundamentally different from the population we may aim to study, for instance small-scale farmers, fishermen or indigenous groups. Differences between population groups can be so large as to make it impossible to argue that findings obtained with student pools are generalizable to these groups, once they have been statistically controlled for socio-economic differences. Additionally, such differences take place at the university entry level, so that selection bias cannot be corrected through stratified sampling.

Finally, and closely related to the above is the fact that including context and realism in the experiment can be *a priori* important for the researcher, and these two elements can make the use of student subjects unfeasible, given their lack of prior experience on the topic. For instance, in a recent study, Alpízar, Carlsson and Johansson (2008a,b) studied prosocial behavior from visitors to a protected area when they were exposed to a design that included hypothetical and real decisions. The aim of comparing a hypothetical situation to a real one, in a real context, clearly in this case invalidated the use of students as the subject pool.

It is important to add that moving away from student subject pools may have disadvantages, especially from the perspective of the interpretation of the experiment. The risk of participants using *ad hoc* rules to answer some questions they might have not understood is greater. It is necessary to be prepared for participants to perceive the experimental environment from a more distant perspective than students, given that their level of education may be low and they are not used to following written instructions. Consequently, it is convenient if the experiment has a low level of complexity, the time allocated for reading instructions and giving examples and practice rounds is adequate and with sufficient clarity, and the design of each field experiment goes through repeated pilot tests before it is finally approved.

Contextualization and type of information

Frequently, the research development agenda in developing countries is highly influenced by the availability of financial sources: foundations, NGOs, government agencies and international development agencies provide funds for experimental research (in many cases even through the form of consulting). This originated from a demand for realism requiring the participation of experimental subjects and of groups of special interest, which includes using realistic conditions in the experimental design. This implies both challenges and opportunities.

- The challenge is to achieve a scientifically solid research program that is generalizable from, sometimes, very specific contexts and hastened work agendas.
- Working at the request of the decision makers allows the formulation of more interesting hypotheses with potentially greater policy relevance.

As an illustration, as researchers we could be interested in studying how members in a community make decisions, in general, regarding the use of a scarce common resource. For instance, Moreno-Sánchez and Maldonado (2010) conducted field experiments with eight fishing communities located in the surroundings of a sea-protected area (SPA). The managers of the area supported the experimental design of this study. The aim was to evaluate a strategy of co-management in the use of marine resources and to compare it to the outcomes resulting from externally imposed regulation. The experiments worked as a pedagogical tool and have been the basis for initiating a discussion of alternative strategies in the use of MPA in these communities.

Such scientific interest may well coincide with the work agenda of an NGO in a specific fishing community facing these types of problems. For the NGO, and for the research program, it is desirable to use members of the community as subject pools, so that they can contribute their experiences and life knowledge when participating in the experiment. Moreover, such experience could be used to its fullest if the experiment had a realistic context and was calibrated with information derived from the state of marine resources in the community.

For example, if the tasks are designed so that they resemble daily decisions of how to use resources by a representative fisherman of the community, and if the payoffs for one given action or another are comparable to the daily earnings of an artisanal-fishery, then the choices made by the fisherman in the experiment could closely resemble his daily decisions, and changes in those decisions would constitute a reaction to the treatments (for instance, changes in the incentives to respect a ban). The better the design and execution of the experiment, the more attributable the reaction could be to the treatment, and the more relevant the findings will be both for science and for the *in situ* management of the fishing resource.

Typically, field experiments in development economics have been contextualized, and such contextualization is frequently an object of study in and of itself. Mostly, the experiment consists of:

i.   Experienced and informed participants.
ii. A realistic context.
iii. They make decisions over goods, services or investments closely adjusted to reality.
iv. Institutions and rules are familiar to participants.
v.   The amounts paid and the form of payment (money, hours worked, rice bags, etc.) reflect the reality of the activity.

Such levels of realism bring various additional challenges to a typical field experiment design. The first is a logistical one: in studying development issues, a field experiment needs to be conducted in marginal or rural areas, without access to sophisticated computer equipment and in often precarious physical locations. In a recent study, Alpízar, Carlsson and Naranjo (2011) organized eleven workshops with small-scale coffee producers, in eleven different community rooms in schools. Each workshop required chairs and tables to be moved, the preparation of coffee and snacks and the organization of parallel activities for the children in the community whose parents were participating in the experiment.

Another challenge is managing the accumulation of experiences that participants bring to an experiment. If the experiment is realistic in its fundamental components, such life experiences become an intrinsic part of the experiment. For instance, inhabitants of areas marked by violence carry in them resentment and lack of trust towards authorities. Therefore the experimenter must pay special attention to not being labeled as a representative of the government, which happens frequently when universities are perceived as official institutions.

Finally, the third is an ethical challenge. This is particularly relevant in a contextualized or natural field experiment. Because the aim of the experiment is to reflect reality, it is necessary to think about the consequences that will follow after the final round. This forces the experimenter to use truthful information that cannot be misunderstood, and to make sure that any behavior arising in the experiment (especially if it is a negative behavior) would not go beyond the experimental environment. For instance, breaking verbal agreements in coordination experiments may end up in recrimination and disputes that can easily go beyond the experimental environment, and which can easily affect a community's social capital.

The path walked by experimental economics in the Latin American context and its influence in the region

Latin American has been a fruitful ground for the use and development of experimental tools and their applications in the field, which is the central topic of this chapter. Mostly, these applications have focused on problems in rural areas and to a great extent on environmental and natural resource problems. By presenting a general view of the studies that have been developed in the region we can identify various studies focused on measuring behavioral aspects such as saving, trust, and social capital (Barr and Packard, 2000; Lazzarini *et al.*, 2004; Schechter 2004, 2007). There are also present in the region anthropological studies using experimental methods (see, for instance, the studies of Henrich *et al.*, 2006, 2010; Kirby *et al.*, 2002). The greater part of these studies in economics, psychology and anthropology have focused on measuring preferences and behavior, with the aim of comparing their findings with those observed in industrialized countries, where such studies are more frequent. As an example of research on conspicuous consumption and relative income in Costa Rica see Alpízar *et al.* (2005).

In another study of two watersheds in Kenya and Colombia, Cárdenas *et al.* (2011) conducted an experiment inspired by a problem in the voluntary provision of public goods, and the extraction of common-use resources, to emulate a cooperation problem between those located on the upper part of an irrigation system and those located downstream. The distribution of resources in the experiment was made sequentially, starting with those upstream and then downstream, imposing an additional sequential externality to the provision problem. In this study the authors observed that lack of trust between those in the upper and bottom part would lead to unequal distributions of resources. This phenomenon was observed in both the experimental location but also in the actual geographical location of the participants.

In another line of work, mainly developed by research from Latin American countries, we can find experiments motivated by problems related to public policy and development problems particular to the region. A good example of this is studying the role that external regulations play in the construction and destruction of behavioral patterns for the sustainable extraction of natural resources. Many of these studies feature regulators and regulations with a limited capacity to observe and sanction rules violations, very much in line with the reality of public policy in natural resources and with regulator agents with weak enforcement capacity and subject to problems of surveillance and sanctions.

In many experimental studies, researchers from the region have explored the interaction between social preferences and external regulations (some examples for Colombia are Cárdenas *et al.*, 2000; López *et al.*, 2012; Moreno-Sánchez and Maldonado, 2010; Vélez *et al.*, 2010). These research lines have allowed the study of complementarities and substitution between social norms and economic regulations with direct applications to the problem of designing adequate instruments to fight poverty or social efficiency in the context of weak state presence in rural communities.

A characteristic common to all these studies is that the environments they are applied to inspire their designs. Many were conducted in natural national parks and in the context of existing or potentially applicable practices and regulations, or are inspired by ecological conditions such as watersheds (Alpízar *et al.*, 2008a, 2008b; Cárdenas *et al.*, 2011; Moreno-Sánchez and Maldonado, 2010). Alpízar and colleagues worked with the agency managing a national park in Costa Rica to study, through an experimental design, different mechanisms for the allocation of entrance fees and donations to the park. The experiment was applied to real visitors to the park aiming to unravel whether a greater willingness to pay above the fees charged by the park would allow improved funding and financing of the protected areas (see also section 2 in Chapter 1).

Other interesting studies have been those where the researcher establishes a collaboration linked to a direct application of a program aimed at reducing poverty and providing micro-financing, such as the case of the FINCA program in Peru (see www.fincaperu.net, Karlan, 2005, 2007).

There are still unexplored areas for research in Latin America and the application of economic experiments. Two of them are prominent:

- The application of experiments to the analysis of how individuals behave in competitive markets and in those where market power generates inefficiency problems.
- The study of behavior relevant to urban contexts: such as, the contribution to the provision of public services, abiding by social norms, corruption problems, and collective action in urban social movements.

These are topics of great relevance in the region, since larger cities keep drawing immigrants from the rural areas, with the problems of rural and urban poverty continuing to impose substantial challenges on the government.

Other examples of research areas where experimental methods can greatly contribute to the design and adjustment of public policies are linked to the role that programs of focalized and conditional cash transfers may play in the general economic and financial behavior of the most vulnerable populations. Programs such as Oportunidades/ Progresa in Mexico, Familias en Accion in Colombia and Bolsa Familia in Brazil have shown positive impacts on the investment in human capital due to subsidies. However, they also pose important questions on the behavior households show in other consumption areas, crowding out issues – in particular, issues of labor supply – in the mid and long term.

It is still early to evaluate the impact that experimental approaches may have on the design of institutional changes and on public policy programs in developing countries. The developed world is just starting to implement concrete applications derived from field and laboratory experiments, such as those developed by Charles Plott, Alvin Roth and Vernon Smith (see Chapter 3 for more details).

An environment that may be open to impact from experimental methods is academia, and in particular undergraduate and graduate programs. However, the region is still in its infancy in terms of building laboratories, as well as in relation to including experimental economics courses at undergraduate and graduate levels, even in the most prestigious programs of the region. There are still few professors with training in this methodological area and curricular structure still reflects the more orthodox versions of economic science. A first attempt at regional integration was started in 2008, aiming to join students and researchers interested in the region with the purpose of exchanging experiences in the use of experimental methods in economics.[3]

There is still a way to go in Latin America for experiments to impact public programs and policies, although we believe that the different experiments presented here, that have directly or indirectly involved local actors, could sow a seed of reflection in the design of local regulatory mechanisms, or of the assumptions that policy makers may have regarding the behavior of economic agents. The works of Alpízar *et al.* (2008 a, b) and Moreno-Sánchez and Maldonado (2010) mentioned above are good examples of this because the employees at a national park were involved in the design of the experiments. However, despite the closeness field experiments have to the realities of institutional design, they have not yet generated an automatic improvement in design.

While programs and policies are designed with the orthodox tendencies of conventional economics, and do not incorporate the advances made by behavioral sciences, it will be very difficult to include the findings and conclusions from experimental research in their design. A good example of an incorrect assumption is that all the users of fishing, forest or irrigation resources behave according to the "tragedy of the commons." Experimental and ethnographic evidence suggests that merely a third or less of the population behaves according to this prediction.

Based on the experience gathered from experimental research we have been able to identify a series of ideal characteristics that would allow an experimental study to have the potential to impact public programs and policies:

- Policy makers and decision makers were involved in experimental design from the beginning, either because they were the original motivators of the questions studied or because they had provided the financial support for the study, thus illustrating the value of dialogue between the scientific questions the researcher asks and the practical concerns of the funding body. The researcher must maintain the rigor of his experimental design in terms of control, sampling, randomness, the types of incentives and payments in the experiment, among others.
- In some cases the experiments can be part of a wider program or project and, in such cases, are complementary to other methods of social measurement and intervention.
- The selection of participants in the experiment is discussed and agreed with representatives of the objective population involved in the research problem and with other actors (*stakeholders*) involved.
- The support and participation of representatives of these agents (for instance, government agencies and NGOs, local organizations and the civil society) was obtained for the execution of these experiments, either as monitors or in the processing, analysis and interpretation of the information.
- The research findings are socialized and discussed with the representatives of those participants and other agents involved.
- The researcher can establish a more durable relationship and reach for carrying out new visits to the locations or communities with whom he conducted the experiments and eventually he can conduct a follow up on participants' behavior, both in the laboratory and daily life.

Ashraf *et al.* (2015) is an ambitious project exploring the financial behavior of Salvadorans, where they build through a random intervention a series of possible mechanisms to determine the control that immigrants can have on the remittances they send to their families and, in particular, over their rates of saving. In addition, the project allows the study of the positive external effects in the saving behavior of the receivers, in circumstances where those remittances are not considered.

A last element, of vital relevance in experimental design with applications to public programs and policies, is that of ethical implications. In a previous section we mentioned the ethical challenge of disseminating information obtained in an experiment and the possible consequences of participants' posterior behavior once the experiment has concluded. When experimental aspects are introduced into public programs or policies additional ethical challenges appear. For instance, random allocation of participants to different treatments can lead to inconveniences arising from inequalities in the distribution of the benefits of the program: especially when it is related to material benefits, such as money and other elements of economic value. In this respect, it is necessary to be completely transparent in the allocation of control groups and treatments and, as much as possible, look for mechanisms for involving the control groups in the future benefits of the program, without sacrificing the rigor of the experimental design.

Another challenge arises when governmental and non-governmental agencies, even for reasons that arise from the legitimate objectives of the program, decide to intervene in the experimental design, sacrificing its experimental rigor; for instance, in the sampling or the application of treatment variables in a way that may induce results in the short run towards those benefiting their organizational mission. The ethical risks can even fortuitously impact programs and interventions, such as the program of financial support to groups of women in Kenya (Gugerty and Kremer, 2008), where an experiment was used to help reduce poverty. The authors show that women with better connections and higher levels of education end up controlling the spaces of associative groups, so that they hindered access to financial resources and finally displaced older women with less education, which further increased the exclusion of the most vulnerable women in the community.

Conclusions

The application of economic experiments in different laboratory and field versions is a reality in developing countries and, although its use is still very recent, the steady increase in the amount and quality of experiments conducted has provided considerable guidance as to the path that should be followed and on what can be learned. There is already an accumulation of experience for the academics that have collaborated with international research networks, governmental and non-governmental agencies in tackling important development problems. Experimental evidence from Latin America has now enriched the study of human behavior in facing problems of environmental conservation, the provision of public goods or the sustainable exploitation of common-use resources. Growing rapidly in number within these countries, is an experimental research agenda that is looking into problems of trust, cooperation, social capital, risk and social networks.

To the extent that these studies have replicated and adapted the methods used in other locations, it has been possible to draw interesting conclusions on development problems in developing countries. Given that replicability is one of the virtues of economic experiments, we can say after a first evaluation that a set of positive and promising results has been observed. Hopefully some of the studies referenced in this chapter have enabled a positive preliminary evaluation of the path that has been followed.

Moreover, we believe that the especial closeness experimentalists in these countries have with the reality of development problems and with international, national and local actors interested in using these tools, has brought additional benefits to enrich the scientific agenda as well as the design of public programs and policies. The permanent dialogue and association with the financial supporters of development programs invites scientists to think of appropriate experimental designs that are applicable and useful to enrich the answers to questions of development. On the other hand, the growing formalization of promotion and evaluation systems in academia in developing countries forces researchers to maintain rigor in their designs and in their contribution to the frontier of knowledge in their corresponding research areas. These two conditions become determinants of future research in terms of the application of experimental methods to development problems.

Notes

1. The reader can find in Cárdenas and Carpenter (2008) an extensive review of experimental studies conducted in developing countries or studies on development problems, which is difficult to summarize here given space considerations.
2. This section is mainly based on an article by Harrison and List (2004). This article, published in the *Journal of Economic Literature*, is an obligatory reference to describe a field experiment.
3. Latin American Field Experiments Network http://www.decon.edu.uy/ LAFEN/

References

Abbink, Klaus, and Jordi Brandts. 2005. "Price Competition under Cost Uncertainty: A Laboratory Analysis." *Economic Inquiry*, 43: 636–648.

Abbink, Klaus, and Jordi Brandts. 2008. "Pricing in Bertrand Competition with Increasing Marginal Costs." *Games and Economic Behavior*, 63: 1–31.

Advani, Raj, Walter Yuan, and Peter Bossaerts. 2003. jMarkets. Free Software.

Akerlof, George A. 1982. "Labor Contracts as Partial Gift Exchange." *The Quarterly Journal of Economics*, 97(4): 543–569.

Akerlof, George A., and Rachel Kranton. 2005. "Identity and the Economics of Organizations." *Journal of Economic Perspectives*, 19: 9–32.

Alchian, Armen A., and Harold Demsetz. 1972. "Production, Information Costs and Economic Organization." *The American Economic Review*, 62(5): 777–795.

Alevy, Jonathan E., Michael S. Haigh, and John A. List. 2007. "Information Cascades: Evidence from a Field Experiment with Financial Market Professionals." *The Journal of Finance*, 62(1): 151–180.

Alpízar, Francisco, Fredrik Carlsson and Olof Johansson. 2005. "How Much Do We Care About Absolute versus Relative Income and Consumption?" *Journal of Economic Behavior & Organization*, 56(3): 405–421.

Alpízar, Francisco, Fredrik Carlsson and Olof Johansson. 2008a. "Anonymity, Reci- procity and Conformity: Evidence from Voluntary Contributions to a National Park in Costa Rica." *Journal of Public Economics*, 92(5–6): 1047–1060.

Alpízar, Francisco, Fredrik Carlsson and Olof Johansson. 2008b. "Does Context Matter More for Hypothetical than for Actual Contributions? Evidence from a Natural Field Experiment." *Experimental Economics*, 11: 299–314.

Alpízar, Francisco, Fredrik Carlsson and Maria A. Naranjo. 2011. "The effect of risk, ambiguity, and coordination on farmers: adaptation to climate change: A Framed Field Experiment." *Ecological Economics,* 70, 2317–2326.

Altavilla, Carlo, Luigi Luini and Patrizia Sbriglia. 2006. "Social Learning in Market Games." *Journal of Economic Behavior & Organization,* 61: 632–652.

Anderhub, Vital, Simon Gächter, and Manfred Königstein. 2002. "Efficient Contracting and Fair Play in a Simple Principal-Agent Experiment." *Experimental Economics*, 5(1): 5–27.

Angerer, Martin, Jurgen Huber and Michael Kirchler. 2009. "Experimental Asset Markets with Endogenous Choice of Costly Asymmetric Information." Unpublished: Available at http://ssrn.com/abstract=1399359.

Aragonés, Enriqueta, and Thomas Palfrey. 2004. "The Effect of Candidate Quality on Electoral Equilibrium: An Experimental Study." *American Political Science Review*, 98: 77–90.

Aragonés, Enriqueta, and Thomas Palfrey. 2005. "Electoral Competition between Two Candidates of Different Quality: The Effects of Candidate

Ideology and Private Information." Social Choice and Strategic Decisions: Essays in Honor of Jeffrey S. Banks, David Austen-Smith and John Duggan, eds. Berlín: Springer.

Arbak, Emrah, and Marie-Claire Villeval. 2007. "Endogenous Leadership: Selection and Influence." *IZA Discussion Paper* No. 2732.

Argueta, Rafael, Diego Aycinena, Edwin Castro, Juan Carlos Córdova, Juan Carlos Moratya, and Fernando Moscoso. 2010. "Use of a Combinatorial Auction to Allocate 6 BOO Power Transmission Contracts: A Case Study about Guatemala." Unpublished.

Ariely, Dan, Uri Gneezy, George Loewenstein and Nina Mazar. 2009. "Large Stakes and Big Mistakes." *Review of Economic Studies*, 76(2): 451–469.

Arifovic, Jasmina, Janet Hua Jiang, and Yiping Xu. 2013. "Experimental Evidence of Bank Runs as Pure Coordination Failures." *Journal of Economic Dynamics and Control*, 37(12): 2446–2465.

Arrow, Kenneth J., and Gerard Debreu. 1954. "Existence of an Equilibrium for a Competitive Economy." *Econometrica*, 22: 265–290.

Ashraf, Nava, Diego Aycinena, Claudia A. Martínez, and Dean Yang. 2015. "Savings in Transnational Households: A Field Experiment among Migrants from El Salvador." *Review of Economics and Statistics*, 97(2): 332–351.

Asparouhova, Elena, Peter Bossaerts, Jon Eguia, and William Zame. (2015). "Asset Pricing and Asymmetric Reasoning." *Journal of Political Economy*, 123(1): 66–122.

Ausubel, Lawrence, Peter Cramton, and Paul Milgrom. 2006. "The Clock proxy Auction: A Practical Combinatorial Auction Design. Combinatorial Auctions." Peter Cramton, Yoav Shoham and Richard Steinberg, eds. Cambridge: MIT Press. 115–138.

Avery, Christopher, and Peter Zemsky. 1998. "Multidimensional Uncertainty and Herd Behavior in Financial Market." *The American Economic Review*, 88(4): 724–748.

Babus, Ana. 2014. "The Formation of Financial Networks." Unpublished.

Baker, George, Robert Gibbons and Kevin J. Murphy. 2001. "Bringing the Market inside the Firm?" *The American Economic Review*, 91(2): 212–218.

Banerjee, Abhijit V. 1992. "A Simple Model of Herd Behavior." *The The Quarterly Journal of Economics*, 107: 797–817.

Banerjee, Abhijit V., and Esther Duflo. 2009. "The Experimental Approach to Development Economics." *Annual Review of Economics*, 1: 151–178.

Barr, Abigail, and Truman Packard. 2000. "Revealed and Concealed Preferences in the Chilean Pension System: An Experimental Investigation" Department of Economics Discussion Paper Series, University of Oxford.

Barreda Tarrazona, Ivan, Aurora García Gallego, Nikolaos Georgantzís, Joaquin Andaluz, and Augustin Gil. 2011. "An Experimental Study of Spatial Competition with Endogenous Pricing." *International Journal of Industrial Organization*, 29: 74–83.

Barro, Robert J., and David B. Gordon. 1983. "A Positive Theory of Monetary Policy in a Natural Rate Model." *Journal of Political Economy*, 91: 589–610.

Battaglini, Marco, Rebecca Morton and Thomas Palfrey. 2010. "The Swing Voter's Curse in the Laboratory." *Review of Economic Studies*, 77: 61–89.

Bergstrom, Theodore C., and John H. Miller. 1999. Experiments with Economic Principles, 2nd edition. New York: McGraw Hill.

Bernanke, Ben S. 1983. "Nonmonetary Effects of the Financial Crisis in the Propagation of the Great Depression." *The American Economic Review*, 73(3): 257–276.

Bertrand, Joseph F.L. 1883. "Thèorie Mathèmatiques de la Richesse Sociale." *Journal des Savants*, 67: 499–508.

Besley, Timothy, and Stephen Coate. 1997. "An Economic Model of Representative Democracy." *The Quarterly Journal of Economics*, 112: 85–114.

Bikhchandani, Sushil, David Hirshleifer, and Ivo Welch. 1992. "A Theory of Fads, Fashion, Custom, and Cultural Change as Informational Cascades." *Journal of Political Economy*, 100(5): 992–1026.

Bolton, Gary, and Axel Ockenfels. 2000. "ERC: A Theory of Equity, Reciprocity and Competition." *The American Economic Review*, 90: 166–193.

Bornstein, Gary, 1992. "The Free-Rider Problem in Intergroup Conflict over Step-Level and Continuous Public Goods." *Journal of Personality and Social Psychology*, 62: 597–606.

Bolton, Patrick, and Mathias Dewatripont. 2005. "Contract Theory." Cambridge: MIT Press.

Bosch-Domènech, Antoni, and Joaquim Silvestre. 1997. "Credit Constraints in General Equilibrium: Experimental Results." *The Economic Journal*, 107(444): 1445–1464.

Bosch-Domènech, Antoni, and Nicolaas Vriend. 2003. "Imitation of Successful Behavior in Cournot Markets." *The Economic Journal*, 113: 495–524.

Bosch-Domènech, Antoni, and Shyam Sunder. 2000. "Tracking the Invisible Hand: Convergence of Double Auctions to Competitive Equilibrium." *Computational Economics*, 16: 257–284.

Bossaerts, Peter, and Charles Plott. 2004. "Basic Principles of Asset Pricing Theory: Evidence from Large-Scale Experimental Financial Markets." *Review of Finance, Springer*, 8(2): 135–169.

Bossaerts, Peter, Charles Plott and William Zame. 2007. "Prices and Portfolio Choices in Financial Markets: Theory, Econometrics, Experiments." *Econometrica*, 75: 993–1038.

Bossaerts, Peter, Paolo Ghirardato, Serena Guarnaschelli and William Zame. 2010. "Ambiguity in Asset Markets: Theory and Experiments." *Review of Financial Studies*, 23: 1325–1359.

Bowles, Samuel. 1998. "Endogenous Preferences: The Cultural Consequences of Markets and other Economic Institutions." *Journal of Economic Literature*, 36: 75–111.

Brandts, Jordi, and David Cooper. 2006. "A Change Would Do You Good: An Experimental Study of How to Overcome Coordination Failure in Organizations." *The American Economic Review*, 96: 669–693.

Brandts, Jordi, and David Cooper. 2007. "It's What You Say Not What You Pay: An Experimental Study of Manager-employee Relationships in Overcoming Coordination Failure." *Journal of the European Economic Association*, 5: 1223–1268.

Brandts, Jordi, David Cooper and Enrique Fatás. 2007. "Leadership and Overcoming Coordination Failure with Asymmetric Costs." *Experimental Economics*, 10: 269–284.

Brandts, Jordi, Paul Pezanis-Christou and Alan Schram. 2008. "Competition with Forward Contracts: A Laboratory Analysis Motivated by Electricity Market Design." *The Economic Journal*, 118(525): 192–214.

Brandts, Jordi, Christina Rott and Carles Solà. 2015. "Not Just like Starting Over: Leadership and Revivification of Cooperation in Groups." Barcelona GSE Working Paper Series, Working Paper no. 775.

Brañas Garza, Pablo. 2006. "Poverty in Dictator Games: Awakening Solidarity." *Journal of Economic Behavior & Organization*, 60: 306–320.

Brañas Garza, Pablo. 2007. "Promoting Helping Behavior with Framing in Dictator Games." *Journal of Economic Psychology*, 28: 477–486.

Brañas Garza, Pablo, Ramón Cobo, María Paz Espinosa, Natalia Jiménez, Jaromir Kovarik and Giovanni Ponti. 2010. "Altruism and Social Integration." *Games and Economic Behavior*, 69: 249–257.

Brañas Garza, Pablo, María Paz Espinosa and Pedro Rey Biel. 2011. "Travelers' Types." *Journal of Economic Behavior & Organization*, 78: 25–36.

Brown, Martin, Stefan T. Trautmann, and Razvan Vlahu. 2014. "Understanding Bank Run Contagion." Unpublished.

Brown-Kruse, Jamie, and David J. Schenk. 2000. "Location, Cooperation and Communication: An Experimental Examination." *International Journal of Industrial Organization*, 18: 59–80.

Cabrales, Antonio, Piero Gottardi and Fernando Vega Redondo. 2014. "Risk Sharing and Contagion in Networks." *Society for Economic Dynamics Meeting Papers* 278.

Cadigan, John. 2005. "The Citizen Candidate Model: An Experimental Analysis." *Public Choice*, 123: 197–216.

Calomiris, Charles W., and Joseph R. Mason. 2003. "Fundamentals, Panics, and Bank distress during the depression." *The American Economic Review*, 93: 1615–1647.

Camacho Cuena, Eva, Aurora García-Gallego, Nikolaos Georgantzís, and Gerardo Sabater Grande. 2005. "Buyer-Seller Interaction in Experimental Spatial Markets." *Regional Science and Urban Economics*, 35: 89–105.

Camera, Gabriele, Charles N. Noussair, and Steven Tucker. 2003. "Rate of Return Dominance and Efficiency in an Experimental Economy." *Economic Theory*, 22: 629–660.

Camerer, Colin, and Ulrike Malmendier. 2007. "Behavioral Economics of Organizations." Behavioral Economics and Its Applications. P. Diamond and H. Vartiainen, eds. Princeton: Princeton University Press.

Camerer, Colin, George Loewenstein and Matthew Rabin. 2003. "Advances in Behavioral Economics." Princeton: Princeton University Press.

Camerer, Colin. 1987. "Do Biases in Probability Judgment Matter in Markets? Experimental Evidence." *The American Economic Review*, 77: 981–997.

Candelo, Carmen, Juan C. Cárdenas, J. E. Correa, María Claudia López, Diana Lucía Maya, María Ximena Zorrilla and Ana María Roldán. 2002. Juegos económicos y diagnóstico rural participativo. *Un manual con ejemplos de aplicación para la cooperación*. Universidad Javeriana and WWF Colombia.

Cantillon, Estelle, and Martin Pesendorfer. 2006. Auctioning Bus Routes: The London Experience. Combinatorial Auctions. Peter Cramton, Yoav Shoham and Richard Steinberg, ed. 115–138. Cambridge: MIT Press.

Cárdenas, Juan C., John K. Stranlund and Cleve E. Willis. 2000. "Local Environmental Control and Institutional Crowding-out." *World Development*, 28(10): 1719–1733.

Cárdenas, Juan C., and Jeffrey Carpenter. 2008. "Behavioral Development Eco- nomics: Lessons from Field Labs in the Developing World." *Journal of Development Studies*, 44(3): 337–364.

Cárdenas, Juan C., Luz Ánsgela Rodríguez and Nancy Johnson. 2011. "Collective Action for Watershed Management: Field Experiments in Colombia and Kenya." *Environment and Development Economics*, 16 (3): 275–303.

Chakravarty, Surajeet, Miguel A. Fonseca and Todd R. Kaplan. 2014. "An Experiment on the Causes of Bank Runs Contagions." *European Economic Review*, 72: 39–51.

Chamberlin, Edward H. 1948. "An Experimental Imperfect Market." *Journal of Political Economy*, 56(2): 95–108.

Charness, Gary, and Peter Kuhn. 2010. "Lab Labor: What Can Labor Economists Learn from the Lab?" *Handbook of Labor Economics*, 4A. Orley Ashenfelter and David Card eds. 229–330. Elsevier North-Holland.

Charness, Gary, and Matthew Rabin. 2002. "Understanding Social Preferences with Simple Tests." *The Quarterly Journal of Economics*, 117: 817–869.

Chen, Yan, and Sherry Li. 2009. "Group Identity and Social Preferences." *The American Economic Review*, 99: 431–457.

Cipriani, Marco, and Antonio Guarino. 2005. "Herd Behavior in a Laboratory Financial Market." *The American Economic Review*, 95(5): 1427–1443.

Cipriani, Marco, and Antonio Guarino. 2008. "Transaction Costs and Informational Cascades in Financial Markets." *Journal of Economic Behavior & Organization*, 68(3): 581–592.

Cipriani, Marco, and Antonio Guarino. 2009. "Herd Behavior in Financial Markets: an Experiment with Financial Market Professionals." *Journal of the European Economic Association*, 7(1): 206–233.

Clark, Kenneth, and Martin Sefton. 2001. "The Sequential Prisoner's Dilemma: Evidence on Reciprocation." *The Economic Journal*, 111: 51–68.

Collier, Kenneth E., Richard D. McKelvey, Peter C. Ordeshook, and Kenneth C. Williams. 1987. "Retrospective Voting: An Experimental Study." *Public Choice*, 53: 101–130.

Collins, Richard, and Katerina Sherstyuk. 2000. "Spatial Competition with Three Firms: An Experimental Study." *Economic Inquiry*, 38: 73–94.

Condorcet, Marquis de. 1785. "Essai sur L'application de L'analyse à la Probabilité des Décisions Rendues à la Pluralité des Voix." París: Imprimerie Royale.

Coppinger, Vicki M., Vernon L. Smith and Jon A. Titus. 1980. "Incentives and Behavior in English, Dutch and Sealed-bid Auctions." *Economic Inquiry*, 18(1): 1–22.

Corbae, Dean, and John Duffy. 2008. "Experiments with Network Formation." *Games and Economic Behavior*, 64(1): 81–120.

Cournot, Augustin A. 1838. "Recherches sur les Principes Mathèmatiques de la Thèorie des Richesses." París: Hachette.

Cramton, Peter, Yoav Shoham and Richard Steinberg. 2006. "Introduction to Combinatorial Auctions." Peter Cramton, Yoav Shoham and Richard Steinberg, eds. 1–13. Cambridge: MIT Press.

Croson, Rachel T. A., Enrique Fatas and Tibor Neugebauer. 2005. "Conditional Cooperation in Two Public Goods Games: The Weakest Link and the VCM." *Economics Letters*, 2005, 87: 97–101.

Croson, Rachel T. A., Enrique Fatas, Tibor Neugebauer and Antonio J. Morales. 2015. "Excludability: A Laboratory Study on Forced Ranking in Team Production." *Journal of Economic Behavior & Organization*, 114: 13–26.

Cyert, Richard, and Morris H. DeGroot. 1973. "An Analysis of Cooperation and Learning in a Duopoly Context." *The American Economic Review*, 63: 26–37.

Davis, Douglas D., and Bart J. Wilson. 2005. "Differentiated Product Competition and the Antitrust Logit Model: An Experimental Analysis." *Journal of Economic Behavior & Organization*, 57: 89–113.

Davis, Douglas D., and Charles A. Holt. 1993. "Experimental Economics." Princeton: Princeton University Press.

Davis, Douglas D., and Robert J. Reilly. 2014. "Looking Tough and Letting Them Know: The Effects of Payment Renegotiation and Information on Financial Fragility," Unpublished.

Davison, Lee K., Carlos D. Ramirez. 2014. "Local Banking Panics of the 1920s: Identification and Determinants." *Journal of Monetary Economics*, 66: 164–177.

Deck, Cary A., Kevin A. McCabe and David P. Porter. 2006. "Why Stable Fiat Money Hyperinflates: Results from an Experimental Economy." *Journal of Economic Behavior & Organization*, 61: 471–486.

Denton, Michael, Stephen Rassenti, Steven Backerman and Vernon L. Smith. 2001. "Market Power in a Deregulated Electrical Industry." *Decision Support System*, 30(3): 357–381.

De Graeve, Ferre and Alexei Karas. 2014. "Evaluating Theories of Bank Runs with Heterogeneity Restrictions." *Journal of the European Economic Association*. 12(4): 969–996.

Diamond, Douglas. W., and Philip H. Dybvig. 1983. "Bank Runs, Deposit Insurance, and Liquidity." *Journal of Political Economy*, 91: 401–419.

Dijk, Oege. 2014. "Bank Run Psychology," Unpublished.

Downs, Anthony. 1957. An Economic Theory of Democracy. New York: Harper & Row.

Drehmann, Mathias, Jörg Oechssler, and Andreas Roider. 2005. "Herding and Contrarian Behavior in Financial Markets: An Internet Experiment." *The American Economic Review*, 95(5): 1403–1426.

Duffy, John. 2014. "Macroeconomics: A Survey of Laboratory Research," Unpublished.

Dufwenberg, Martin. 2014. "Banking on Experiments?" Bocconi University, Working Paper no. 534.

Durham, Yvonne. 2000. "An Experimental Examination of Double Marginalization and Vertical Relationships." *Journal of Economic Behavior & Organization*, 42: 207–230.

Eckel, Catherine C., and Philip J. Grossman. 2005. "Managing Diversity by Creating Team Identity." *Journal of Economic Behavior & Organization*, 58(3): 371–392.

Edgeworth, Francis Y. 1881. Mathematical Psychics. London: Kegam Paul.

Ellsberg, Daniel. 1961. "Risk, Ambiguity, and the Savage Axioms.." *The Quarterly Journal of Economics*, 75: 643–669.

Engel, C. 2007. "How much collusion? A meta-analysis of oligopoly experiments." *Journal of Competition Law and Economics*, 3: 491–549.

Ennis, Huberto. M. 2003. "Economic Fundamentals and Bank Runs." *FRB Richmond Economic Quarterly*, 89(2): 55–71.

Epstein, Rafael, Lysette Henríquez, Jaime Catalán, Gabriel Y. Weintraub and Cristián Martínez. 2002. "A Combinatorial Auction Improves School Meals in Chile." *Interfaces*, 32(6): 1–14.

Erev, Ido, and Amnon Rapoport. 1990. "Provision of Step Level Public Goods: The Sequential Contribution Mechanism." *Journal of Conflict Resolution*, 34(3): 401–425.

Exadaktylos, Filippos, Antonio Espín, and Pablo Brañas-Garza. 2013."Experimental Subjects are not different." *Scientific Reports*, 3: 1213, 1–6.

Fatas, Enrique, Antonio J. Morales, and Paloma Úbeda. 2010. "Blind Justice." *Journal of Economic Psychology*, 31(3): 358–373.

Feddersen, Timothy, and Wolfgang Pesendorfer. 1996. "The Swing Voter's Curse." *The American Economic Review*, 86: 408–424.

Fehr, Ernst, Georg Kirchsteiger, and Arno Riedl. 1993. "Does Fairness Prevent Market Clearing? An Experimental Investigation." *The Quarterly Journal of Economics*, 108: 437–460.

Fehr, Ernst, Erich Kirchler, Andreas Weichbold, and Simon Gächter. 1998. "When Social Norms Overpower Competition: Gift Exchange in Experimental Labor Markets." *Journal of Labor Economics*, 16(2): 324–351.

Fehr, Ernst, and Jean-Robert Tyran. 2001. "Does Money Illusion Matter?" *The American Economic Review*, 91: 1239–1262.

Fehr, Ernst, and Jean-Robert Tyran. 2005. "Individual Irrationality and Aggregate Outcomes." *Journal of Economic Perspectives*, 19: 43–66.

Fehr, Ernst, and Jean-Robert Tyran. 2007. "Money Illusion and Coordination Failure." *Games and Economic Behavior*, 58: 246–268.

Fehr, Ernst, and Jean-Robert Tyran. 2008. "Limited Rationality and Strategic Interaction: The Impact of the Strategic Environment on Nominal Inertia." *Econometrica*, 76: 353–394.

Fehr, Ernst, and Klaus M. Schmidt. 1999. "A Theory of Fairness, Competition and Cooperation." *The Quarterly Journal of Economics*, 114: 817–868.

Fehr, Ernst, and Klaus M. Schmidt. 2006. "The Economics of Fairness, Reciprocity and Altruism – Experimental Evidence and New Theories." In Kolm, S-C, Ythier, J. M. (Eds.), *Handbook of the Economics of Giving, Altruism and Reciprocity*, 1: 615–691.

Ferreira, Jose Luis, Praveen Kujal, and Stephen Rassenti. 2010. "Multiple Openings of Forward Markets: Experimental Evidence." *Economics Working Papers*, we1023. Universidad Carlos III.

Fershtman, Chaim, and Kenneth L. Judd. 1987. "Equilibrium Incentives in Oligopoly." *The American Economic Review*, 77: 927–940.

Fischbacher, Urs. 2007. "z-Tree: Zurich Toolbox for Readymade Economic Experiments." *Experimental Economics*, 10: 171–178.

Fisher, Eric O'N. 2001. "Purchasing Power Parity and Interest Parity in the Laboratory." *Australian Economic Papers*, 40: 586–602.

Fisher, Joseph, Mark Isaac, Jeffrey Schatzberg and James Walker. 1995. "Heterogeneous Demand for Public Goods: Behavior in the Voluntary Contributions Mechanism." *Public Choice*, 85: 249–266.

Ford, Jefrey, and Laurie Ford. 1995. "The Role of Conversation in Producing Intentional Change in Organizations." *Academy of Management Review*, 20: 541–570.

Fonseca, Miguel A., and Hans-Theo Normann. 2008. "Mergers, Asymmetries and Collusion: Experimental Evidence." *The Economic Journal*, 118: 387–400.

Fonseca, Miguel A., and Hans-Theo Normann. 2012. "Explicit vs. Tacit Collusion – The Impact of Communication in Oligopoly Experiments." *European Economic Review*, 56: 1759–1772.

Fonseca, Miguel A., Steffen Huck and Hans-Theo Normann. 2005. "Playing Cournot Although They Shouldn't." *Economic Theory*, 25: 669–677.

Fonseca, Miguel A., Wieland Müller and Hans-Theo Normann. 2006. "Endogenous Timing with Observable Delay in Duopoly: Experimental Evidence." *International Journal of Game Theory*, 34: 443–456.

Forsythe, Robert, and Russell Lundholm. 1990. "Information Aggregation in an Experimental Market." *Econometrica*, 58: 309–347.

Forsythe, Robert, Thomas Reitz and Robert Weber. 1993. "An Experiment on Coordination in Multi-Candidate Elections: The Importance of Polls and Election Histories." *Social Choice and Welfare*, 10: 223–247.

Fouraker, Lawrence, and Sidney Siegel. 1963. *Bargaining Behavior*. New York: McGraw Hill.

Gächter, Simon, and Falk, A. 2002. "Reputation and Reciprocity: Consequences for the Labor Relation." *Scandinavian Journal of Economics*, 104: 1–26.

Gächter, Simon, Daniele Nosenzo, Elke Renner, and Martin Sefton. 2012. "Who Makes a Good Leader? Cooperativeness, Optimism and Leading by Ex- ample." *Economic Inquiry*, 50(4): 953–967.

García-Gallego, Aurora. 1998. "Oligopoly Experimentation of Learning with Simulated Markets." *Journal of Economic Behavior & Organization*, 35: 333–335.

García-Gallego, Aurora, and Nikolaos Georgantzís. 2001a. "Multiproduct Activity in an Experimental Differentiated Oligopoly." *International Journal of Industrial Organization*, 19: 493–518.

García-Gallego, Aurora, and Nikolaos Georgantzís. 2001b. "Adaptive Learning by Single-Product and Multiproduct Price-Setting Firms in Experimental Markets." *Instituto Valenciano de Investigaciones Económicas WP-AD 2001–2013*.

García-Galligo, Aurora, Nikolaos Georgantzís, and Gerardo Sabater Grande. 2004. "Identified Consumers: An Experiment on the Informativeness of Cross-Demand Price Effects." *Cuadernos de Economía*, 27: 185–216.

García-Gallego, Aurora, Nikolaos Georgantzís, and Praven Kujal. 2008. "Experimental Insights into the Efficiency of Alternative Water Management Institutions." *Game Theory and Policy Making in Natural Resources and the Environment*. Dinar, A., J. Albiac and J. Sánchez Soriano, eds. 209–235. London/New York: Routledge, Taylor & Francis Group.

Garratt, Rod, and Todd Keister. 2009. "Bank Runs as Coordination Failures: An experimental study." *Journal of Economic Behavior & Organization*, 71(2): 300–317.

Georgantzís, Nikolaos, Constantine Manasakis, Evaggelos Mitrokostas, and Emanuel Petrakis. 2008. "Strategic Delegation in Experimental Duopolies with Endogenous Incentive Contracts." University of Crete, Department of Economics, Working Paper no. 0809.

Gneezy, Uri, and Aldo Rustichini. 2000. "Pay Enough or Don't Pay At All." *The Quarterly Journal of Economics* 115(3): 791–810.

Gode, Dhananjay K., and Shyam Sunder. 1993a. "Allocative Efficiency of Markets with Zero intelligence Traders: Market as a Partial Substitute for Individual Rationality." *Journal of Political Economy*, 101(1): 119–137.

Gode, Dhananjay K., and Shyam Sunder. 1993b. "Lower Bounds for Efficiency of Surplus Extraction in Double Auctions." The Double Auction Market. Santa Fe Institute Studies in the Sciences of Complexity, Proc. Volume XIV. New York: Addison Wesley.

Goldstein, Itay, and Ady Pauzner. 2005. "Demand–deposit Contracts and the Probability of Bank Runs." *The Journal of Finance*, 60(3): 1293–1327.

Goodfellow, Jessica, and Charles R. Plott. 1990. "An Experimental Examination of the Simultaneous Determination of Input Prices and Output Prices." *Southern Economic Journal*, 56(4): 969–983.

Gorton, Gary. 1988. "Banking Panics and Business Cycles." *Oxford Economic Papers*, 40: 751–781.

Gorton, Gary, and Andrew Winton. 2003. "Financial intermediation." *Handbook of the Economics of Finance*, 1: 431–552.

Gu, Chao. 2011. "Herding and Bank Runs." *Journal of Economic Theory*, 146(1): 163–188.

Guarnaschelli, Serena, Richard McKelvey, and Thomas Palfrey. 2000. "An Experimental Study of Jury Decision Rules." *American Political Science Review*, 94: 407–423.

Gugerty, Mary Kay and Michael Kremer. 2008. "Outside Funding and the Dynamics of Participation in Community Organizations" *American Journal of PoliticalScience*, 52(3): 585–602.

Harrison, Glenn, and John A. List. 2004. "Field Experiments." *Journal of Economic Literature*, XLII: 1009–1055.

Harstad, Ronald, Stephen Martin, and Hans-Theo Normann. 1998. "Intertemporal Pricing Schemes: Experimental Tests for Consciously Parallel Behavior in Oligopoly." *Applied Industrial Economics*. L. Philips, ed. 123–151. Cambridge: Cambridge University Press.

Harstad, Ronald. 2000. "Dominant Strategy Adoption and Bidders Experience with Pricing Rules." *Experimental Economics*, 3(3): 261–280.

Hart, Oliver, and Bengt Holmstrom. 1987. "The Theory of Contracts." *Advances in Economic Theory, Fifth World Congress*. Truman Bewley ed. Cambridge: Cambridge University Press, 71–155.

Hawking, Stephen. 2001. "The Universe in a Nutshell." Cambridge University Press.

Heinemann, Frank. 2012. "Understanding Financial Crises: The contribution of experimental economics." *Annals of Economics and Statistics*, 7–29.

Henrich, Joseph, Richard McElreath, Abigail Barr, Jean Ensminger, Clark Barrett, Alexander Bolyanatz, Juan C. Cárdenas, Michael Gurven, Edwins Gwako, Natalie Henrich, Carolyn Lesorogol, Frank Marlowe, David Tracer, and John Ziker. 2006. "Costly Punishment Across Human Societies." *Science*, 312: 1767–1770.

Henrich, Joseph, Jean Ensimger, Richard McElreath, Abigail Barr, Clark Barrett, Alexander Bolyanatz, Juan C. Cárdenas, Michael Gurven, Edwins Gwako, Natalie Henrich, Carolyn Lesorogol, Frank Marlowe, David Tracer, and John Ziker. 2010. "Markets, Religion, Community Size, and the Evolution of Fairness and Punishment." *Science*, 327: 1480–1484.

Hens, Thorsten, Klaus R. Schenk-Hoppe, and Bodo Vogt. 2007. "The Great Capitol Hill Baby Sitting Co-op: Anecdote or Evidence for the Optimum Quantity of Money?" *Journal of Money, Credit and Banking*, 39: 1305–1333.

Hey, John D., and Daniela di Cagno. 1998. "Sequential Markets: An Experimental Investigation of Clower's Dual Decision Hypothesis." *Experimental Economics*, 1(1): 63–85.

Hinloopen, Jeroen, Wieland Müller, and Hans-Theo Normann. 2014, "Output Commitment through Product Bundling – Experimental Evidence." *European Economic Review*, 65: 164–180.

Holmstrom, Bengt, and Jean Tirole. 1989. "The Theory of the Firm." *Handbook of Industrial Economics Part 1*. R. Schmalensee, and R.Willig, comp. Amsterdam: Elsevier Publishing B.V.

Hölmstrom, Bengt. 1982. "Moral Hazard in Teams." *Bell Journal of Economics*, 13: 324–340.

Holt, Charles A. 1995. "Industrial Organization: A Survey of Laboratory Research." *Handbook of Experimental Economics*. Kagel, John, and Alvin Roth, eds. 349–444. Princeton: Princeton University Press.

Holt, Charles A., Loren W. Langan, and Anne P. Villamil. 1986. "Market Power in Oral Double Auctions." *Economic Inquiry*, 24(1): 107–123.

Hong, James T., and Charles R. Plott. 1982. "Rate Filing Policies for Inland Water Transportation: An Experimental Approach." *Bell Journal of Economics*, 13: 1–19.

Hotelling, Harold. 1929. "Stability in Competition." *The Economic Journal*, 39: 41–57.

Huck, Steffen, Hans-Theo Normann, and Jorg Oechssler. 1999. "Learning in Cournot Oligopoly: An Experiment." *The Economic Journal*, 109: 80–95.

Huck, Steffen, Hans-Theo Normann, and Jorg Oechssler. 2000. "Does Information About Competitors Actions' Increase or Decrease Competition in Experimental Oligopoly Markets?" *International Journal of Industrial Organization*, 18: 39–57.

Huck, Steffen, Hans-Theo Normann, and Jorg Oechssler. 2001b. "Market Volatility and Inequality in Earnings: Experimental Evidence." *Economics Letters*, 70: 363–368.

Huck, Steffen, Hans-Theo Normann, and Jorg Oechssler. 2004b. "Two are few and four are Many: Number Effects in Experimental Oligopolies." *Journal of Economic Behavior and Organization*, 53: 435–446.

Huck, Steffen, Wieland Müller, and Vicky Knoblauch. 2006. "Spatial Voting with Endogenous Timing." *Journal of Institutional and Theoretical Economics*, 162: 557–570.

Huck, Steffen, Wieland Müller, Kai Konrad, and Hans-Theo Normann. 2007. "The Merger Paradox and Why Aspiration Levels Let it Fail in the Laboratory." *The Economic Journal*, 522: 1073–1095.

Huck, Steffen, Wieland Müller, and Hans-Theo Normann. 2001a. "Stackelberg Beats Cournot on Collusion and Efficiency in Experimental Markets." *The Economic Journal*, 111: 749–766.

Huck, Steffen, Wieland Müller, and Hans-Theo Normann. 2002a. "To Commit or Not to Commit: Endogenous Timing in Experimental Duopoly Markets." *Games and Economic Behavior*, 38: 240–264.

Huck, Steffen, Wieland Müller, and Hans-Theo Normann. 2004a. "Strategic Delegation in Experimental Markets." *International Journal of Industrial Organization*, 22: 561–574.

Huck, Steffen, Wieland Müller, and Nicolaas J. Vriend. 2002b. "The East End, the West End, and King's Cross: On Clustering in the Four Player Hotelling Game." *Economic Inquiry*, 40: 231–240.

Hurwicz, Leonid, Roy Radner, and Stanley Reiter. 1975a. "A Stochastic Decentralized Resource Allocation Process: Part 1." *Econometrica*, 43: 187–221.

Hurwicz, Leonid, Roy Radner, and Stanley Reiter. 1975b. "A Stochastic Decentralized Resource Allocation Process: Part 2." *Econometrica*, 43: 363–393.

Ichniowski, Casey, Kathryn Shaw, and Giovanna Prennushi. 1997. "The Effects of Human Resource Management Practices on Productivity: A Study of Steel Finishing Lines." *The American Economic Review*, 87: 291–313.

Isaac, R. M., and James Walker. 1991. Costly Communication: An Experiment in a Nested Public Goods Problem, Contemporary Laboratory Research in Political Economy. Thomas Palfrey, comp. Ann Arbor: University of Michigan Press.

Isaac, R. M., and James M. Walker. 1988a. "Group Size Effects in Public Goods Provision: The Voluntary Contribution Mechanism." *The Quarterly Journal of Economics*, 103(1): 179–199.

Isaac, R.M., and James M. Walker. 1988b. "Communication and Free Riding Be- havior: The Voluntary Contributions Mechanism." *Economic Inquiry*, 26: 585–608.

Iyer, Rajkamal, and Manju Puri. 2012. "Understanding Bank Runs: The Importance of Depositor-Bank Relationships and Networks." *The American Economic Review*, 102(4): 1414–1445.

Kagel, John H., and Dan Levin. 1985. "Individual Bidder Behavior in First Price Private Value Auctions." *Economic Letters*, 19: 125–128.

Kagel, John H., and David Levin. 1995–2008. "Auctions: A Survey of Experimental Research." *Handbook of Experimental Economics*. John H. Kagel, and Alvin E. Roth, ed. Princeton: Princeton University Press.

Kagel, John H., and David Levin. 1993. "Independent Private Value Auctions: Bidder Behaviour in First, Second and Third price Auctions with Varying Numbers of Bidders." *The Economic Journal*, 103(419): 868–879.

Kagel, John H., Ronald Harstad, and David Levin. 1987. "Information Impact and Allocation Rules in Auctions with Affiliated Private Values: A Laboratory Study." *Econometrica*, 55(6): 1275–1304.

Kagel, John H., 1995. Auctions: "A Survey of Experimental Research." *Handbook of Experimental Economics*. John H. Kagel, and Alvin E. Roth, eds. Princeton: Princeton University Press. 501–586.

Kandel, Eugene, and Edward Lazear. 1992. "Peer Pressure and Partnerships." *Journal of Political Economy*, 100: 801–817.

Karlan, Dean. 2005. "Using Experimental Economics to Measure Social Capital and Predict Financial Decisions." *The American Economic Review*, 95(5): 1688–1699.

Karlan, Dean. 2007. "Social Connections and Group Banking." *The Economic Journal*, 117: F52–F84.

Kelly, Frank. 1995. "Laboratory Subjects as Multiproduct Monopoly Firms: An Experimental Investigation." *Journal of Economic Behavior & Organization*, 27: 401–420.

Kelly, Morgan, and Cormac Ó Gráda. 2000. "Market Contagion: Evidence from the Panics of 1854 and 1857." *The American Economic Review*, 90(5): 1110–1124.

Kirby, Kris N., Ricardo Godoy, Victoria Reyes García, Elizabeth Byron, Lilian Apaza, William Leonard, Eddy Pérez, Vincent Vadez, and David Wilkie. 2002. "Correlates of Delay-discount Rates: Evidence from Tsimane Amerindians of the Bolivian Rain Forest." *Journal of Economic Psychology*, 23: 291–316.

Kiss, Hubert Janos, Ismael Rodriguez-Lara, and Alfonso Rosa-García. 2012. "On the Effects of Deposit Insurance and Observability on Bank Runs: An Experimental Study." *Journal of Money, Credit and Banking*, 44(8): 1651–1665.

Kiss, Hubert Janos, Ismael Rodriguez-Lara, and Alfonso Rosa-García. 2014a. "Do Social Networks Prevent or Promote Bank Runs?" *Journal of Economic Behavior & Organization*, 101: 87–99.

Kiss, Hubert Janos, Ismael Rodriguez-Lara, and Alfonso Rosa-García. 2014b. "Do Women Panic more than Men? An Experimental Study of Financial Decisions." *Journal of Behavioral and Experimental Economics*, 52: 40–51.

Kiss, Hubert Janos, Ismael Rodriguez-Lara, and Alfonso Rosa-García. 2015. "Think Twice Before Running! Bank Runs and Cognitive Abilities." *Forthcoming in Journal of Behavioral and Experimental Economics*.

Klemperer, Paul. 2002. "What Really Matters in Auction Design." *Journal of Economics Perspectives*, 16(1): 169–189.

Klos, Alexander, and Norbert Sträter. 2013. "How Strongly Do Players React to Increased Risk Sharing in an Experimental Bank Run Game." *QBER Discussion Paper*.

Knez, Marc, and Duncan Simester. 2002. "Form Wide Incentives and Mutual Monitoring At Continental Airlines." *Journal of Labor Economics*, 19: 743–772.

Kotter, John. 1996. "Leading Change." Boston: Harvard University School Press.

Kremer, Michael. 1993. "The O-Ring Theory of Economic Development." *The Quarterly Journal of Economics*, 107: 551–575.

Kreps, David. 1996. "Corporate Culture and Economic Theory. Firms, Organizations and Contracts: A Reader in Industrial Organization." P. J. Buckley, and J. Michie, eds. Oxford University Press.

Kydland, Finn E., and Edward C. Prescott. 1977. "Rules Rather than Discretion: The Inconsistency of Optimal Plans." *Journal of Political Economy*, 85: 473–490.

Kübler, Dorothea, and Wieland Müller. 2002. "Simultaneous and Sequential Price Competition on Heterogeneous Duopoly Markets: Experimental Evidence." *International Journal of Industrial Organization*, 20: 1437–1460.

Ladha, Krishna, Gary Miller, and Joe Oppenheimer. 2003. "Information Aggregation by Majority Rule: Theory and Experiments." http://www.bsos. umd.edu/gvpt/oppenheimer/research/jury.pdf.

Lazzarini, Sergio, Regina Madalozzo, Rinaldo Artes, and José de Oliveira Siqueira. 2004. "Measuring Trust: An Experiment in Brazil." IBMEC Working Paper- WPE-2004–1.

Le Coq, Chloé, and Henrik Orzen. 2006. "Do Forward Markets Enhance Competition?: Experimental Evidence." *Journal of Economic Behavior & Organization*, 61(3): 415–431.

Ledyard, John O., Mark Olson, David Porter, Joseph A. Swanson, and David P. Torma. 2002. "The First Use of a Combined Value Auction for Transportation Services." *Interfaces*, 32(5): 4–12.

Ledyard, J. 1995. "Public Goods: A Survey of Experimental Research." Handbook of Experimental Economics. J. Kagel, and A. E. Roth, eds. Princeton: Princeton University Press.

Lei, Vivian, Charles, N. and Charles Plott. 2001. "Non-Speculative Bubbles in Experimental Asset Markets: Lack of Common Knowledge of Rationality vs. Actual Irrationality." *Econometrica*, 69: 831–859.

Levine, David K., and Thomas Palfrey. 2007. "The Paradox of Voter Participation? A Laboratory Study." *American Political Science Review*, 101: 143–158.

Lian, Peng, and Charles R. Plott. 1998. "General Equilibrium, Markets, Macroeconomics and Money in a Laboratory Experimental Environment." *Economic Theory*, 12(1): 21–75.

List, John A., and David Lucking-Reiley. 2002. "Demand Reduction in Multiunit Auctions: Evidence from a Sportscard Field Experiment." *The American Economic Review*, 90(4): 961–972.

López, María Claudia, James J. Murphy, John M. Spraggon, and John K. Stranlund 2012. "Comparing the Effectiveness of Regulation and Pro-Social Emotions to Enhance Cooperation: Experimental Evidence from Fishing Communities in Colombia." *Economic Inquiry*, 50(4): 131–142.

Lucking-Reiley, David. 2000. "Vickery Auctions in Practice: From Nineteenth century Philately to Twenty first century E-commerce." *Journal of Economic Perspectives*, 14(3): 183–192.

Macho-Stadler, Inés and David Pérez-Castrillo. 2001. *An Introduction to the Economics of Information: Incentives and Contracts* (2nd edition), Oxford: Oxford University Press.

Makowski, Louis. 1980. "A Characterization of Perfectly Competitive Economies with Production." *Journal of Economic Theory*, 22(2): 208–221.

Marshall, Alfred. 1890. "Principles of Economics: Volume I." London: Macmillan.

Martin, S., Hans-Theo Normann, and C.M. Snyder. 2001. "Vertical Foreclosure in Experimental Markets." *Rand Journal of Economics*, 3: 466–496.

Mason, Charles F., and Owen R. Phillips. 1997. "Information and Cost Asymmetry in Experimental Duopoly Markets." *Review of Economics and Statistics*, 49: 290–299.

Maximiano, Sandra, Randolph Sloof, and Joep Sonnemans. 2007. "Gift Exchange in a Multi-worker Firm." *The Economic Journal*, 117(522): 1025–1050.

McCabe, Kevin A. 1989. "Fiat Money as a Store of Value in an Experimental Market." *Journal of Economic Behavior & Organization*, 12: 215–231.

McCabe, Kevin A., Stephen J. Rassenti, and Vernon L. Smith. 1991a. "Smart Computer assisted Markets." *Science*, 254(5031): 534–534.

McCabe, Kevin A., Stephen J. Rassenti, and Vernon L. Smith. 1991b. "Testing Vickrey's and Other Simultaneous Multiple Unit Versions of the English Auction." *Research in Experimental Economics*, Vol. 4. R. M. Isaac, ed. Greenwich: JAI Press. 4: 45–79.

McKelvey, Richard D., and Peter C. Ordeshook. 1982. "Two-Candidate Elections without Majority Rule Equilibria: An Experimental Study." *Simulation and Games*, 13: 311–335.

McKelvey, Richard D., and Peter C. Ordeshook. 1990. "A Decade of Experimental Research on Spatial Models." Advances in the Spatial Theory of Voting. James M. Enelow, and Melvin J. Hinich, eds. Cambridge: Cambridge University Press.

Moreno-Sánchez, Rocío, and Jorge Maldonado. 2010. "Evaluating the Role of Co-management in Improving Governance of Marine Protected Areas: An Ex- perimental Approach in the Colombian Caribbean." *Ecological Economics*, 69: 2557–2567.

Morris, Stephen, and Hyun Song Shin. 2003. "Global Games: Theory and Applications. Advances in Economics and Econometrics." (Proceedings of the Eighth World Congress of the Econometric Society), edited by M. Dewatripont, L. Hansen and S. Turnovsky; Cambridge University Press.

Müller, Wieland. 2006. "Allowing for Two Production Periods in the Cournot Duopoly: Experimental Evidence." *Journal of Economic Behavior & Organization*, 60: 100–111.

Nagel, Rosemarie, and Nicollas J. Vriend. 1999. "An Experimental Study of Adaptive Behavior in an Oligopolistic Market Game." *Journal of Evolutionary Economics*, 9: 27–65.

Nalbantian, Haig R., and Andrew Schotter. 1997. "Productivity under Group Incentives: An Experimental Study." *The American Economic Review*, 87(3): 314–341.

Normann, Hans-Theo. 2002. "Endogenous Timing with Incomplete Information and with Observable Delay." *Games and Economic Behavior*, 39: 282–291.

Normann, Hans-Theo. 2011. "Vertical Mergers, Foreclosure And Raising Rivals' Costs: Experimental Evidence." *Journal of Industrial Economics*, 59: 506–527.

Noussair, Charles N., Charles R. Plott, and Raymond G. Riezman. 1995. "An Experimental Investigation of the Patterns of International Trade." *The American Economic Review*, 85: 462–491.

Noussair, Charles N., Charles R. Plott, and Raymond G. Riezman. 1997. "The Principles of Exchange Rate Determination in an International Financial Experiment." *Journal of Political Economy*, 105: 822–861.

Noussair, Charles N., Charles R. Plott, and Raymond G. Riezman. 2007. "Production, Trade, Prices, Exchange Rates and Equilibration in Large Experimental Economies." *European Economic Review*, 51: 49–76.

Offerman, Theo, Jans Potters, and Joep Sonnemans. 2002. "Imitation and Belief Learning in an Oligopoly Experiment." *Review of Economic Studies*, 69: 973–997.

Offerman, Theo, and Joep Sonnemans. 1998. "Learning by Experience and Learning by Imitating Successful Others." *Journal of Economic Behavior & Organization*, 34: 559–576.

Olson, Mark A., Stephen J. Rassenti, Vernon L. Smith, Mary L. Rigdon, and Michael J. Ziegler. 1999. "Market Design and Motivated Human Trading Behavior in Electricity Markets." Proceedings of the 32nd International Conference on Systems Sciences.

Orzen, Henrik. 2008. "Counterintuitive Number Effects in Experimental Oligopolies." *Experimental Economics*, 11: 390–401.

Osborne, Martin, and Al Slivinski. 1996. "A Model of Political Competition with Citizen Candidates." *The Quarterly Journal of Economics*, 111: 65–96.

Ostroy, Joseph M. 1980. "The No-surplus Condition as a Characterization of Perfectly Competitive Equilibrium." *Journal of Economic Theory*, 22(2): 183–207.

Palfrey, Thomas R. 2009. "Laboratory Experiments in Political Economy" *AnnualReview of Political Science*, 12: 379–388.

Palfrey, Thomas R., and Howard Rosenthal. 1985. "Voter Turnout with Strategic Uncertainty." *Public Choice*, 41: 7–53.

Plott, Charles R. 1982. "Industrial Organization Theory and Experimental Economics." *Journal of Economic Literature*, 20 (4): 1485–1527.

Plott, Charles R., and Vernon L. Smith. 1978. "An Experimental Examination of Two Exchange Institutions." *Review of Economic Studies*, 45: 133–153.

Plott, Charles R., and Shyam Sunder. 1982. "Efficiency of Experimental Security Markets with Insider Information: An Application of Rational Expectations Models." *Journal of Political Economy*, 90: 663–698.

Plott, Charles R. 1991. "A Comparative Analysis of Direct Democracy, Two Candidate Elections, and Three Candidate Elections in and Experimental Environment." *Laboratory Research in Political Economy*. Thomas Palfrey, ed. Ann Arbor: University of Michigan Press.

Porter, David, and Stephen J. Rassenti. 2010. "Combinatorial Auctions." Wiley Encyclopedia of Operations.

Porter, David, Stephen J. Rassenti, Anil Roopnarine, and Vernon L. Smith. 2003. "Combinatorial Auction Design." *Proceedings of the National Academy of Sciences*, 100(19): 11153.

Porter, David, and Vernon L. Smith. 2006. "FCC License Auction Design: A 12 year Experiment." *Journal of Law, Economics and Policy*, 3: 63–85.

Potters, Jan, and Sigrid Suetens. 2013. "Oligopoly Experiments in the Current Millennium." *Journal of Economic Surveys*, 27, 439–460.

Prendergast, Canice. 1999. "The Provision of Incentives in Firms." *Journal of Economic Literature*, 37(1): 7–63.

Rabin, Matthew. 1993. "Incorporating Fairness into Economics and Game Theory." *The American Economic Review*, 83: 1281–1302.

Rassenti, Stephen J., Stanley S. Reynolds, Vernon L. Smith, and Ferenc Szidarovszky. 2000a. "Adaptation and Convergence of Behavior in Repeated Experimental Cournot Games." *Journal of Economic Behavior & Organization*, 41: 117–146.

Rassenti, Stephen J., Vernon L. Smith, and Bart J. Wilson. 2003. "Controlling Market Power and Price Spikes in Electricity Networks: Demand side Bidding." Proceedings of the *National Academy of Sciences*, 100(5): 2998.

Rassenti, Stephen J., Vernon L. Smith, and Bart J. Wilson. 2003. "Discriminatory Price Auctions in Electricity Markets: Low Volatility at the Expense of High Price Levels." *Journal of Regulatory Economics*, 23(2): 109–123.

Rassenti, Stephen J., Vernon L. Smith, and Robert L. Bulfin. 1982. "A Combinatorial Auction Mechanism for Airport Time Slot Allocation." *Bell Journal of Economics*, 13(2): 402–417.

Reiley, David H., Sai-Ming Li, and Randall A. Lewis. 2010. "Northern Exposure: A Field Experiment Measuring Externalities Between Search Advertisements." Proceedings of the *11th ACM Conference on Electronic Commerce*. David C. Parkes, Chrysant-hos Dellarocas, and Moshe Tennenholtz, ed., 297–304.

Ricardo, David. 1817. "On the Principles of Political Economy, and Taxation." London: John Murray.

Riedl, Arno and Frans van Winden. 2001. "Does the Wage Tax System cause Budget Deficits?" *Public Choice*, 109: 371–394.

Riedl, Arno, and Frans van Winden. 2007. "An Experimental Investigation of Wage Taxation and Unemployment in Closed and Open Economies." *European Economic Review*, 51: 871–900.

Riker, William, and Peter Ordeshook. 1968. "A Theory of the Calculus of Voting." *American Political Science Review*, 62: 25–42.

Riley, Matt. 2010. "The Rational Optimist." London: Fourth State.

River, Charles and Associates Inc., and Market Design Inc. 1998a. "Report 1B: Package Bidding for Spectrum Licenses." *Charles River and Associates* Report No. 1351-00. Cambridge: Charles River and Associates.

River, Charles and Associates Inc., and Market Design Inc. 1998b. "Report 2: Simultaneous Ascending Auctions with Package Bidding." *Charles River and Associates* Report No. 1351-00. Cambridge: Charles River and Associates.

Robbins, Stephen. 2010. "Comportamiento Organizativo." Prentice Hall.

Rotemberg, Julio. 1994. "Human Relations in the Workplace." *Journal of Political Economy*, 102(4): 684–717.

Roth, Alvin E. 2002. "The Economist as Engineer: Game Theory Experimentation and Computation as Tools for Design Economics." *Econometrica*, 70(4): 1341–1378.

Roth, Alvin E., and Tayfun Sönmez. 2005. "A Kidney Exchange Clearinghouse in New England." *The American Economic Review*, 95(2): 376–380.

Roth, Alvin E., and Tayfun Sönmez. 2007. "Efficient Kidney Exchange: Coincidence of Wants in Markets with Compatibility-based Preferences." *The American Economic Review*, 97(3): 828–851.

Sabater-Grande, Gerardo, and Nikolaos Georgantzís. 2002. "Accounting for Risk Aversion in Repeated Prisoners' Dilemma Games." *Journal of Economic Behavior & Organization*, 48: 37–50.

Salanié, Bernard. 1997. "The Economics of Contracts: A Primer." Cambridge: MIT Press.

Salas, Vicente. 1996. "Economía de la Empresa." Ariel Economía.

Samuelson, Paul A. 1958. "An Exact Consumption loan Model of Interest with or without the Social Contrivance of Money." *Journal of Political Economy*, 66: 467–468.

Schechter, Laura. 2004. "Traditional Trust Measurement and the Risk Confound: An Experiment in Rural Paraguay." *Journal of Economic Behavior & Organization*, 62(2): 272–292.

Schechter, Laura. 2007. "Theft, Gift-Giving, and Trustworthiness: Honesty is Its Own Reward in Rural Paraguay." *The American Economic Review*, 97(5): 1560–1582.

Schotter, Andrew, and Tanju Yorulmazer. 2009. "On the Dynamics and Severity of Bank Runs: An Experimental Study." *Journal of Financial Intermediation*, 18(2): 217–241.

Schramm, Arthur, and Joep Sonnemans. 1996a. "Voter Turnout as a Participation Game." *International Journal of Game Theory*, 25: 385–406.

Schramm, Arthur, and Joep Sonnemans. 1996b. "Why People Vote: Experimental Evidence." *Journal of Economic Psychology*, 17: 417–442.

Shafir, Eldar, Peter Diamond, and Amos Tversky. 1997. "Money Illusion." *The Quarterly Journal of Economics*, 112: 341–374.

Smith, Adam. 1776. "The Wealth of Nations." London: W. Strahan & T. Cadell.

Smith, Vernon L. 1962. "An Experimental Study of Competitive Market Behavior." *Journal of Political Economy*, 70: 111–137.

Smith, Vernon L. 1964. "The Effect of Market Organization on Competitive Equilibrium." *The Quarterly Journal of Economics*, 78: 181–201.

Smith, Vernon L. 1976. "Experimental Economics: Induced Value Theory." *The American Economic Review*, 66(2): 274–279.

Smith, Vernon L. 1982. "Microeconomic Systems as an Experimental Science." *The American Economic Review*, 72(5): 923–955.

Smith, Vernon L. 1989. "Theory, Experiment and Economics." *Journal of Economic Perspectives*, 3(1): 151–169.

Smith, Vernon L., Gerry Suchanek, and Arlington Williams. 1988. "Bubbles, Crashes, and Endogenous Expectations in Experimental Spot Asset Markets." *Econometrica*, 56: 1119–1151.

Smith, Vernon L. 1991. "Rational Choice: The Contrast between Economics and Psychology." *Journal of Political Economy*, 99: 877–897.

Solà, Carles. 2002. "The Sequential Prisoners' Dilemma Game: Reciprocity and Group Size Effects." *Experimental Economics: Financial Markets, Auctions, and Decision Making*. F. Andersson, and H.J. Holm, eds. Kluwer Academic Publishers.

Sprague, Oliver Mitchell Wentworth. 1910. "History of Crises under the National Banking System." US Government Printing Office, Vol. 538.

Starr, Martha A., and Rasim Yilmaz. 2007. "Bank Runs in Emerging Market Economies: Evidence from Turkey's Special Finance Houses." *Southern Economic Journal*, 73: 1112–1132.

Suetens, Sigrid, and Jan Potters. 2007. "Bertrand Colludes more than Cournot." *Experimental Economics*, 10: 71–77.

Sunder, Shyam. 1992. "Market for Information: Experimental Evidence." *Econometrica*, 60: 667–695.

Sweeney, Joan, and Richard J. Sweeney. 1977. "Monetary Theory and the Great Capitol Hill Baby Sitting Co-op Crisis." *Journal of Money, Credit and Banking*, 9: 86–89.

Swenson, Charles W. 1988. "Taxpayer Behavior in Response to Taxation: An Experimental Analysis." *Journal of Accounting and Public Policy*, 7: 1–28.

Trevino, Isabel. 2013. "Channels of Financial Contagion: Theory and Experiments,." Unpublished.

Tziralis, Georgios, and Ilias Tatsiopoulos. 2007. "Prediction Markets: An Extended Literature Review." *Journal of Prediction Markets*, 1(1): 75–91.

Van der Heijden, Eline C. M., Jan H. M. Nelissen, Jan J. M. Potters, and Harrie A. A. Verbon. 1998. "Transfers and the Effect of Monitoring in an Overlapping Generations Experiment." *European Economic Review*, 42: 1363–1391.

Van Dijk, Frans, Joep Sonnemans, and Frans van Winden. 2001. "Incentive Systems in a Real Effort Experiment." *European Economic Review*, 45: 187–214.

Van Huyck, John B., Raymond C. Battalio, and Mary F. Walters. 1995. "Commitment Versus Discretion in the Peasant Dictator Game." *Games and Economic Behavior*, 10: 143–171.

Van Huyck, John B., Raymond C. Battalio, and Mary F. Walters. 2001. "Is Reputation a Substitute for Commitment in the Peasant Dictator Game?" Working Paper, Texas A&M University.

Van Huyck, John, Raymond Battalio, and Richard Beil. 1990. "Tacit Coordination Games, Strategic Uncertainty, and Coordination Failure." *The American Economic Review*, 80(1): 234–248.

Van Koten, Silvester, and Andreas Ortmann. 2010. "Structural Versus Behavioral Remedies in the Deregulation of Electricity Markets: An Experimental Investigation Guided by Theory and Policy Concerns." Working paper, CERGE-EI and EUI Loyola de Palacio.

Vandergrift, Donald, and Abdullah Yavas. 2011. "An Experimental Test of Behavior under Team Production." *Managerial and Decision Economics*, 32(1): 35–51.

Varian, Hal R. 2007. "Position Auctions." *International Journal of Industrial Organization*, 25(6): 1163–1178.

Varian, Hal R. 2009. "Online Ad Auctions." *The American Economic Review*, 99(2): 430–434.

Vélez, María Alejandra, James J. Murphy, and John K. Stranlund. 2010. "Centralized and Decentralized Management of Local Common Pool Resources in the Developing World: Experimental Evidence from Fishing Communities in Colombia." *Economic Inquiry*, 48(2): 254–265.

Vickers, John. 1985. "Delegation and the Theory of the Firm." *The Economic Journal*, 95: 138–147.

Vickery, William. 1962. "Auctions and Bidding Games." *Recent Advances in Game Theory*, 29: 15–27.

Vickrey, William. 1961. "Counter-Speculation, Auctions and Competitive Sealed Tenders." *Journal of Finance*, 16(1): 8–37.

Von Stackelberg, H. 1934. "Marktform und Gleichgewicht." Berlín: Julius Springer.

Walras, Léon. 1874. *Élements d'économie politique pure*. Lausanne: L. Corbaz.

Wessen, Randii R., and David Porter. 2007. "The Cassini Resource Exchange."ASK Magazine, *The Academy Project Engineering Leadership*, 28: 14–18.

Wessen, Randii R., and David Porter. 1998. "Market Based Approaches for Controlling Space Mission Costs: The Cassini Resource Exchange." Journal of Reduced Mission Operations Costs, March 1998.

Wicker, Elmus. 2000. "The Banking Panics of the Great Depression." Cambridge University Press.

Williams, Arlington. 1980. "Computerized Double Auction Markets: Some Initial Experimental Results." *Journal of Business*, 53: 235–258.

Williams, Kenneth. 1991. "Candidate Convergence and Information Costs in Spatial Elections: An Experiment Analysis." Laboratory Research in Political Economy. Thomas Palfrey, ed. Ann Arbor: University of Michigan Press.

Wolfers, Justin, and Eric Zitzewitz. 2004. "Prediction Markets." *Journal of Economic Perspectives*, 18(2): 107–126.

World Bank. 2002. *World Development Indicators 2002*. Washington, DC.

Index

Lightning Source UK Ltd.
Milton Keynes UK
UKOW06n2303101115

262489UK00003B/7/P